D0392472

Designer Food

Designer Food

Mutant Harvest or Breadbasket of the World?

GREGORY E. PENCE

ROWMAN & LITTLEFIELD PUBLISHERS, INC.
Lanham • Boulder • New York • Oxford

ROWMAN & LITTLEFIELD PUBLISHERS, INC.

Published in the United States of America
by Rowman & Littlefield Publishers, Inc.
4720 Boston Way, Lanham, MD 20706
www.rowmanlittlefield.com

12 Hid's Copse Road
Cumnor Hill, Oxford OX2 9JJ, England

British Library Cataloguing-in-Publication Information Available

Library of Congress Cataloging-in-Publication Data

Pence, Gregory E.
Designer food : mutant harvest or breadbasket of the world? / Gregory E. Pence.
p. cm.
Includes bibliographical references (p.) and index.
ISBN 0-7425-0839-0 (cloth : alk. paper)
1. Genetically modified fods—Popular works. I. Title.

TP248.65.F66 P46 2002
363.19'29—dc21

2001041926

Printed in the United States of America

∞™ The paper used in this publication meets the minimum requirements of American National Standard for Information Sciences—Permanence of Paper for Printed Library Materials, ANSI/NISO Z39.48-1992.

Contents

Preface

Food is the very stuff of life. I am what I eat and what I eat is me. In eating, I relate to other people and I relate to the world. How I eat, with whom I eat, and why I eat partly define me. Food is the mind-body problem made personal.

Our current, reductionist ideology tells us that genes make up our true selves; so if we eat genetically modified food, don't we risk changing our natures? When food crops are genetically modified, scientists seem to be changing the very stuff of our lives and, indirectly, to be changing us. Isn't this scary?

Genetically modified (GM) food also scares people because it suggests tampering with our traditional food, which is assumed to be safe, clean, and mostly natural. Growing GM crops suggests the death of the wholesome countryside, which in the past nurtured many virtues and gave many people meaning in life.

The main topics of this book are GM food, whether it should scare us, and whether it is good for the world. When I started writing about GM food, I realized that I could not do so without discussing organic food, world hunger, agricultural terrorists, the safety of meat, environmental ethics, mad cow disease, and European versus American attitudes toward food. At some point, all these topics enter the debate about GM food.

Like death, food concerns everyone. Unlike death, which only happens once and which many people prefer not to think about, food must be considered several times a day, every day, and many people spend a great deal of time thinking about it. Thus it is surprising that philosophers have not written about food, for as we shall see, not only GM food but food policy in general abounds with ethical issues.

Thinking about food can be sharpened by using four worldviews: globalism, naturalism, scientific progressivism, and egalitarianism. I deploy these "frames" throughout the book to illustrate how different people in good faith come to differing conclusions about GM food. For some people, particular frames matter the most in judging GM food. For others, parts of several frames must be combined to make the best judgment.

More is at stake here than aesthetics. Squeamish American and European sensitivities do not mirror the whole world, where many people each night still go to bed hungry. If genetically enhanced crops can feed the starving, prevent blindness, or carry vaccines against pandemics, then these great goods should count vastly in the acceptability of GM crops.

As we shall see, Europeans and Americans react disparately to GM food. Of course, what Europeans fear is McCafés on every block playing American rap music. Can we imagine how we would feel if, say, Japanese or Arabic culture seemed to be triumphing over our own? (Canadian readers undoubtedly *do* understand this feeling.)

At the same time, some European organizations, especially Greenpeace International, inappropriately alarm the public about GM crops. Countering their biases requires some base for comparing the safety of GM foods. In this book, I take that base to be the safety of the meat most Americans eat everyday.

This book also answers some other questions about GM food: is "Bt corn" safe to eat? Does it destroy Monarch butterflies? Does it need smaller amounts of pesticides to grow? Are environmental groups that fight GM food evil? When did past environmentalism assist fascism? Are GM crops safe for the environment? What are their downstream effects?

Extremism characterizes most debates about GM food: egalitarian critics who never see any good in agribusiness fight scientists who view environmentalists as tree-hugging morons; globalists who view small-is-beautiful naturalists as Luddites oppose organic farmers who see proponents of global agriculture as nothing short of industrial Darth Vaders.

Within such polarization, an impartial bioethicist may find employment, at least in removing the conceptual rubbish, finding the common ground, and sorting out the arguments. At least, that is my hope.

Acknowledgments

Two years ago, in June 2000, I was privileged to attend a scenario workshop sponsored by the World Council for Sustainable Development in St. Louis, Missouri. The topic was genetically modified (GM) food and whether the fears gripping Europe over this issue could spread to North America. Talking to the European and Asian journalists, scientists, and politicians there, I came away convinced that it could; thus I started research for this book.

I gave a paper expressing some of the ideas in this book at the 2000 meeting of the American Society for Bioethics in Salt Lake City. I received several valuable suggestions from members of the audience after that talk.

Very early on, I was helped by Mary Ellen Sanders, who owns her own consulting company, Dairy and Food Culture Technologies, based in Denver, Colorado. She is also visiting research scientist at the Dairy Products Technology Center at California Polytechnic State University, San Luis Obispo. Professor Sanders lent me ten years of her files relating to GM food and food safety. She also answered the occasional esoteric question.

Later, I discovered that C. S. Prakash, whose help in my research proved to be invaluable, lived not far from me in Alabama, a hundred miles south in Tuskegee. Prakash (as his friends call him) is the world's most famous and ardent defender of GM food, and is a food scientist at the institution founded by Booker T. Washington. Although we have never met, we have corresponded via e-mail. More important, Professor Prakash told me about his AgBioWorld (a.k.a. "AgView Bio") listserv, an astonishing resource for a writer such as me, where dozens of the top food scientists on the planet once debated, defended, and attacked GM food in over a thousand posted messages (some running for ten pages or more) over two years. I mostly lurked on this listserv, eagerly absorbing the wisdom of this august group. I am sure that I owe a great deal to this group and I apologize if I have taken any of the cases or ideas in this book from this listserv without citing the source. After a while, one forgets where one first read something.

In my research, I was very much helped during the summer of 2000 by my extraordinary research student, Joyce Hsu, who assisted on many of these chapters, tracked down many factual claims, and showed me that she could find almost anything on the Internet. A very smart and friendly person, she was a great coparticipant in this project.

Randolphe H. Wicker of New York City, who takes a keen interest in genetically modified foods, new drugs, and cloning, kindly sent me numerous articles about GM food that he came across. Norman Borlaug, the Nobel Prize winner, was kind enough to encourage me after Joyce Hsu and I wrote an op-ed in the Sunday *Birmingham News* (14 July 200) praising his work on ending starvation. A follow-up op-ed by me ran nationally after Dr. Borlaug was praised by "President Bartlett" on the television show *The West Wing.* I am indebted to him for this and for other support he has given me over the years.

Louis Pojman, philosophy professor at the United States Military Academy, was kind enough to be a reviewer for the book on short notice, as was C. S. Prakash of Tuskegee University. Minnie Randle, the best secretary our philosophy department has had in my twenty-five years at the University of Alabama at Birmingham, also did great work in helping me to produce this book.

1

Organic versus Genetically Modified Food

You don't have to sacrifice style or comfort for an organic lifestyle. It's all about making choices that make a difference in the world—making it cleaner, greener, and healthier.

—*Maria Rodale, editor,* Organic Style *magazine and author,* The Organic Suburbanite

In the modern world, no foods could be more evaluatively distinct than genetically modified (GM) food and organic food. GM food is unnatural, dangerous, and grown by international conglomerates; organic food is natural, safe, wholesome, and grown locally on small farms. GM food departs dangerously from traditional practices, whereas organic food improves on tainted, industrial norms by returning to purer, simpler ways of cultivating food. At least, these are our myths.

In truth, organic food can be unsafe. Organic lettuce or spinach, commonly grown in soil to which manure has been added, can contain *E. coli* bacteria, which can cause hemorrhagic colitis, acute kidney failure, and even death. The Centers for Disease Control estimated that food-borne illnesses caused by the O157:H7 strain of *E. coli* sickened over 73,480 Americans in 1999.[1] Although the etiology of this strain of *E. coli* is poorly understood, in 1993 it was known to have infected more than seven hundred people in western states, killing four, because of undercooked hamburgers served by the Jack-in-the-Box fast-food chain.[2] Children are especially vulnerable to it.

Ironically, eating organic food may expose people to more, not fewer, risks from O157:H7. Outbreaks of *E. coli* O157:H7 stem not only from bad beef, but also from fresh fruit juices, raw milk, lettuce, and minimally processed produce. Unpasteurized apple juice in the Pacific Northwest killed a child and sent sixty-six others to the hospital in the fall of 1996.[3] In 2000, the Tesco supermarket chain pulled from its shelves all organic mushrooms because they tested positive for *E. coli* O157:H7. Putting the lie to the myth that all organic food is local, Tesco said that the suspect batch came from a supplier in Northern Ireland that used Belgian manure.[4]

Understandably, the Beef Industry Food Safety Council wants the public to understand that beef is not the major culprit here. Indirectly referring to the outbreak at the Jack-in-the-Box franchises, the council emphasizes:

> While outbreaks attributed to beef grab headlines, perhaps more attention should be given to fruits and vegetables. The CDC report on surveillance of foodborne disease outbreaks, also released in March 2000, showed that the number of cases of foodborne-illnesses in outbreaks attributed to fruits and vegetables have exceeded those of beef every single year, even by 10-to-1, as in 1995.[5]

So how does organic food become infected with *E. coli*? First, organic food is usually grown in soil with manure, a natural breeding ground for various forms of *E. coli*, as well as any other chemical that might be passed through an animal's body. Although it is a bit gross to dig too deeply into the nature of this pile, one wonders how the assumption got accepted that food grown in manured soil is pure while food that is grown with genes that help plants fix nitrogen from the soil is bad.

Second, growing organic food is labor intensive (Prince Charles's organic garden requires a dozen full-time gardeners) and commercial growers employ unskilled workers to reduce costs. Yuppies who eat organic food now do so because of hundreds of thousands of low-paid workers toiling in back-breaking jobs. Nothing about hand weeding or retilling rows to avoid weeds is easy. Natural ways of growing food are usually primitive and require enormous amounts of human capital.

Ridding organic produce of manure also requires careful washing and such work is boring. But the safety of this food rests on the care of this work. In one case of infected organic produce, bacteria contaminated the water used to rid the produce of manure.[6]

Prima facie, the concepts of organic food and genetically modified food would seem to occupy different universes. Don't these different kinds of food involve groups of plants that share nothing in common? People who believe this, in truth, oversimplify.

Plants that exist in nature generate their own chemicals for killing unwanted insects. Consider rotenone, which comes from the roots and leaves of tropical plants such as jewel vine, derris, and the hoary pea. Commonly sold as a white powder, this potent stuff kills fire ants and is used by organic farmers on tomatoes, pears, and apples. Horticulturists spray it on roses and African violets. Because this pesticide occurs inside a plant that evolved naturally in evolution, it is considered permissible to use it in nearly seven hundred products used in organic gardens.

Scientists at Emory University in Atlanta discovered in 2000 that rotenone injected into rats caused Parkinson's disease–like symptoms.[7] They already knew that some environmental toxin was at work here because farmworkers were seven times more likely than normal to develop Parkinson's.

I am not arguing here that the dangers of organic foods surpass those of GM foods. Rather, I want to expose the naïve assumption that a pesticide (rotenone) created internally by exotic, tropical plants and then manufactured in great quantities and sprayed on traditional food crops is somehow more safe, more benevolent, and less artificial than plants to which a few benevolent genes have been added to the existing thirty thousand.

I do not believe that everything called "natural" is safe and good. Cassava, a natural crop grown in most countries in Africa, when improperly prepared causes acute cassava poisoning, TAN (tropical ataxic neuropathy), Konzo (upper motor neuron paraparesis), and goiter.[8]

Besides, and in a more general sense, who decides what is "natural"? In almost every way, what we commonly call natural is a social construction—in other words, a judgment at a particular time and by a particular culture. In the national parks in North America, the asphalt roads, firefighters who fight lightning fires, forest rangers, pest management, and suppression of dangers to humans like mountain lions, bears, and wolves don't really reflect nature. Many of the plants we take as natural to North America, such as Japanese honeysuckle and purple loosestrife, came as imports in the eighteenth or nineteenth centuries.

Critics admit that millions of Americans have already eaten products from genetically modified corn and soybeans, and while they foresee botanical Armageddon from such modifications, no North American has even been sickened (much less killed!) by eating such GM veggies. We take our system of producing meat as normal and safe, yet it may be far more dangerous than any GM foods.

It is neither splitting hairs nor asking rhetorical questions to argue that GM veggies and organic crops resemble each other more than they differ. Take the infamous Bt corn, alleged to endanger the Monarch butterfly. *Bacillus thuringiensis*, a.k.a. "Bt," occurs naturally in soils around the world. Consumers Union calls this bacterium, a natural pesticide, "more benign than many synthetic [chemical] pesticides."[9] It produces a crystalline protein (a "cry" protein) that kills pests on farms, especially the European corn borer. (Note that this corn borer is a close relative of a caterpillar that becomes the Monarch butterfly, so the U.S. Department of Agriculture and the Food and Drug Administration (FDA) knew that Bt pollen could affect such larvae, especially when artificially dumped on larvae in artificial conditions, but then again, so could Bt as a concentrated spray.)

Because Bt occurs everywhere in soil, and because it does not damage good insects (honeybees, ladybugs) that help control bad ones, it can be sprayed on crops and still have the crops labeled "organic." However, this Bt pesticide must be sprayed and resprayed on crops to achieve the desired effects.

Enter GM corn. Bt corn is nothing more than traditional corn containing inside it the genes of *Bacillus thuringiensis.* In other words, if a plant has been soaked with Bt spray over and over for months, and hence absorbs Bt into its cells, it can still be called "organic," but if we insert Bt genes into the corn and no spray is used, then the corn becomes dangerous, as it's been "genetically modified."

Obviously, organic food in this case does not differ as much from genetically modified food as associations of organic farmers would urge us to believe. No longer conducted only on small farms, organic farming now mimics industrial farming, with some plants imported from huge overseas farms and processed together in huge factories. In the making of what the organic industry calls "an organic TV dinner" (if that is not an oxymoron, what is?), "broccoli is trucked to Alberta, Canada, . . . to meet up with pieces of organic chicken that have already made a stop at a processing plant in Salem, Oregon, where they were defrosted, injected with marinade, cooked, and refrozen."[10]

Because consumers turn to organic foods when they become scared about conventional food, organic growers profit from scares about genetically modified food; e.g., sales of their produce in England soared fivefold during the five years of scares about mad cow disease.[11] Some defenders of traditional agriculture see a well-orchestrated campaign by the natural food industry to scare consumers about GM foods.

At least we know that organic food is better for the world's environment, right? Not so. Indeed, just the opposite. Chemical fertilizers that utilize nitrogen in the atmosphere cannot be used in organic farming, which instead must use animal manure for crops that don't fix their own nitrogen. To fertilize organic crops, animal and human waste, mainly the manure of cattle, must be used. A major supplier to England and Europe, Icelandic Incorporated, destroys forests and grasslands in Ecuador to grow more organic food.

Nobel Prize–winning plant biologist Norman Borlaug laughs at the proposal that the world can be fed on organic foods:

> At the present time, approximately 80 million tons of nitrogen nutrients are utilized each year. If you tried to produce this nitrogen organically, you would require an additional 5 or 6 billion head of cattle to supply the manure. How much wild land would you have to sacrifice just to produce the forage for these cows? There's a lot of nonsense going on here. . . .
>
> As far as plants are concerned, they can't tell whether that nitrate ion comes from artificial chemicals or from decomposed organic matter. If some consumers believe that it's better from the point of view of their health to have organic food, God bless them. Let them buy it. Let them pay a bit more. It's a free society. But don't tell the world that we can feed the present population without chemical fertilizer. That's when this misinformation becomes destructive.[12]

But surely eating organic is more nutritional, right? To quote Borlaug again, "If people want to believe that the organic food has better nutritive value, it's up to them to make that foolish decision. But there's absolutely no research that shows that organic foods provide better nutrition."

Simple-minded assumptions about nature, food, health, and the environment underlie people's embrace of organic foods. Chef Jim White, in his "Kitchen Journal" column in the *Albuquerque Journal*, says, "As I've always said, if you want to be healthy, you have to stay as close to nature as possible. When an animal or plant is raised in an organic environment, you are giving your body what Mother Nature intended it to have."[13]

Chef Jim capitalizes "Mother Nature" to indicate his romantic naturalism. An invisible, Divine Hand guided agriculture of the early nineteenth century, such that the farms of Ohio in 1840 were perfect. If only we could get back to that hardscrabble life, he seems to think, we would all live to be a hundred.

Britain's royal family also debates organic and GM foods. Prince Charles, a critic of GM food, champions organic farming, claiming that it aligns with a vague deity whom Charles variously calls the "Sustainer" or "Creator."[14] Such a deity, Charles assures us, frowns on genetically modified food crops. To grow GM food "usurps our place in Nature." Presumably, growing organic crops does not, but that also implies that tedious hand weeding and spreading of cow manure is our natural place.

Prince Phillip and Princess Anne support genetically modified crops, accusing Charles of "oversimplifications." As Prince Phillip corrected his son publicly, "Do not forget that we have been genetically modifying animals and plants ever since people started selective breeding. People are worried about genetically modified organisms getting into the environment. What people forget is that the introduction of exotic species—like, for instance, the gray squirrel into this country—is going to, or has done, far more damage than a genetically modified piece of potato."[15]

Most scientists agree here with Prince Phillip and Princess Anne. Thousands of plants have been created by being crossbred in traditional ways in which genes randomly mix. Such plants (seedless grapes, tangelos) enter our environment all the time with little testing or regulation. Newly imported foods are also introduced in our grocery stores and restaurants (Kiwi fruit, sushi).

Unless some reason arises to think that a newly introduced food has a protein to which humans are allergic, it is not subjected to the same exhaustive battery of tests that GM veggies undergo. GM veggies get tested more than some of the new food we eat in restaurants or some of the exotic fruits and vegetables that suddenly appear in our grocery stores.

Overall, our present system for testing food is more than a bit hypocritical about the safety of traditional food and more than a bit hysterical about genetically modified veggies. As always, people fear the new kid on the block, especially if his name is "Gene."

END COMMENT

The remaining chapters of this book go into much more detail about GM food and other parts of our modern food system, including our system of bringing meat to the table and some conceptual schemata for thinking about food. In this chapter we have taken the ship out of the safe harbor into the rough seas; now the journey to our destination begins.

2

The Politics of Genetically Modified Food

Americans eat garbage food, they're fat, and they don't know how to eat properly.

—*Pierre Lellouch, a member of France's Parliament*

During the late 1990s, when Englishmen started dying of mad cow disease, Germany's Green Party began attacking what it called "genetically modified organisms" or GMOs. At the time, Greenpeace International began switching from confronting whaling boats to destroying field trials of GMOs (Greenpeace International is the umbrella organization for various national Greenpeace organizations, such as Greenpeace USA). The Green parties of Europe work closely with Greenpeace International and are synonymous with opposition to biotechnology. Since the late 1990s, Greenpeace International has consistently worked the better-safe-than-sorry attitude, arguing that although no evidence exists of dangers to digestion or environment from GMOs, it's better to wait and see before allowing their growth.

Since 1992, branches of the Green Party in Austria, Germany, and Italy have fought against import of genetically modified (GM) crops. Because of their efforts, the European Commission in 1998 voted 402 to 2 to ban import of unlabeled GM food.[1]

Because the Nazis killed more than six million Jews, Gypsies, and homosexuals to foster their genetics of racial hygiene, and because the most important physicians and scientists in German medicine led this racist movement, many Europeans today associate "eugenic" and "genetic" with the arrogant scientific attitudes that led to these deaths. Because of this legacy, Europeans recoil against proposals for genetic enhancement of humans. Genetic therapies for human disease or genetically enhanced crops to resist diseases of plants are unlikely to be tried first in Europe.

Given this history, it is not surprising that phrases such as "genetic engineering" or "genetically modified food" alarm many Germans and Europeans. U.S. secretary of agriculture Dan Glickman underestimated this legacy and how it bolstered European opposition to GM food. As one critic of his performance in Europe wrote:

> Glickman may be missing the deep historical precedent that underlies distrust of genetic science when he admonished Europeans for their "blind adherence to culture and history." Considering the mass annihilation of whole peoples based on faulty genetic science, it is understandable if Europeans view a "master race" of food crops skeptically.[2]

More ominously, the discovery of fatal vCJD—human "mad cow disease"—in young English adults occurred during 1996, precisely when American companies such as Monsanto were slipping GM food into the European market, as they had done in America. Europeans feared this stealth introduction of new genetic food by American companies, and these fears piggybacked on deeper fears of the growing influence of American-style companies on European life.

Greenpeace International talked about the 1996 American corn crop in Europe "as if it were the Normandy invasion."[3] Europeans bought $305 million worth of American corn in 1996, but because of fears of GM food, only $1 million in 1999. After American GM corn and GM soybeans were dumped into the sea, English intellectuals bragged of a "reverse Tea Party."

American companies such as Monsanto misunderstood Europeans' reaction to this new food and responded only by talking the progressive, globalist language of science, safety, and world trade. That was like arguing against creationists by citing facts about evolution. The two sides talked past each other, not addressing the cultural fears that drove the conflict.

Indeed, Americans made matters worse. Secretary Glickman in 1999 threatened to *force* import of American soybeans and corn by raising the club of retaliatory tariffs on imported European food.[4] A speaker for American food growers allegedly once told a group of British/European grocery chains that, regarding GM food, "European objections are irrelevant."[5]

Such threats outraged Europeans, whose anti-Americanism had been growing for years. A Belgian professor opined, "Everybody knows the strawberries and asparagus in America look beautiful but have no flavor."[6] Sadly, this view has some truth. After fifty years of selective breeding for color, size, and thicker skin, the Red Delicious apple grown in the American Northwest has lost most of its taste. As one grower lamented, "Nobody should feel sorry for us—we did this to ourselves [in breeding apples] for color and size and not for taste."[7] (As a result, apple farmers lost $760 million in three years.) And who has not purchased a great-looking tomato or strawberry only to be disappointed at its lack of taste?

European news, as well as American movies, paint a bad picture, showing Americans tolerating urban sprawls and long commutes because of a lack of public transportation, while experiencing polluted air, violence in the inner cities, and racism. Immature, puritanical Americans condemn sex in public life (as shown in "l'affaire Monica") and embrace fundamentalist Christianity[8]—not people sophisticated Europeans want at their dinner parties in Provence.

On other fronts, European youth upset traditionalists by eagerly embracing American movies, music, styles of dress, and computers. Snobbish critics hated EuroDisney when it opened in 1993, fearing greater Americanization would follow. What was next? Las Vegas casinos outside Windsor Castle? Neon billboards on the slopes of the Alps?

Guardians of European high culture saw American food innovations as destroying the great European food traditions. They scorned American inventions such as chewing gum, frosted cornflakes, Twinkies, and Velveeta.

All in all, Europeans increasingly believed that American interests did not coincide with theirs. Two-thirds of Italians and French thought nothing in American culture should guide their own, and 63 percent in 2000 did not feel close to Americans.[9]

Europeans had watched American companies downsize during the early 1980s, putting middle-aged managers out of jobs, a process banned in countries such as France and Germany, where workers enjoy great job security. A member of the French parliament wrote a book devoted to the ugliness he saw in America. As one reviewer summarized his attitude, "It [America] has a record number of armed citizens. It embraces the death penalty, turns the poor away when they need medical care, and its legislators have failed to approve a nuclear test ban."[10]

After the show of force of the American military in the Balkans in 1999, Europeans realized that their meager troops could do nothing in NATO without American agreement. Humiliated, Europeans believed that American values and power ran the planet, and soon thereafter European bookstores carried best-sellers with titles such as *American Totalitarianism, Who Is Killing France? The American Strategy,* and *The World Is Not Merchandise.* Military power plus an unstoppable economy made American culture seem like a tidal wave about to wash away everything distinctively European.

As a result, Europeans saw globalization as an American force, not as an international force in which they participated. As part of this view, they saw GM food as an unnatural, globalist phenomenon pushed by American-type international firms. (Even the unfortunate phrase "GM food," with which we seem to be stuck, sounds like a big American corporation.)

When the World Trade Organization pushed introduction of GM food into Europe, Europeans predictably saw the World Trade Organization (WTO) as a tool of American interests. And perhaps they were correct, for the WTO backed American appeals in twenty-five of twenty-seven cases.

In 2000, when they had the chance, representatives from European nations voted the United States off the UN Human Rights Commission. This snub was more than just an expression of irritation at America's refusal to pay its UN dues, or at the American refusal to ratify the Kyoto treaty to reduce greenhouse gases. It also expressed a genuine hostility to the perceived arrogance of America in world affairs, a situation that grew worse in European eyes when the (more worldly) Clinton left office and the parochial, ignorant (as perceived by Europeans) George Bush entered.

Perhaps Europeans' fears of our culture are not simply irrational and biased, but based on truth. Not everyone's idea of food paradise is barbecue, french fries, and Budweisers.

We Americans *do* live in a hurried culture that has pioneered fast food, TV dinners, the Internet, and drive-thrus. Americans may work more efficiently, but is efficient work the meaning of life? Many middle-aged Americans would probably rather work less and spend more time with their families, and enjoy long family dinners on the weekends.

More specifically, European critics asked, is the way Americans *relate to food* better than our way? Compared to Americans, Europeans characteristically spend more time each day shopping for, preparing, and consuming food. Even if the average American makes more money and can buy more merchandise, does she enjoy her daily meal as much as her European counterpart? Should she miss long lunches in al fresco cafés with native wines? Long morning breaks reading the newspaper at a coffee bar on a public square? Shouldn't enjoyment of good food figure into the good life?

VANDALISM AGAINST FIELD TRIALS OF GM CROPS

The attitudes toward America described above, and the perception that GM crops were American products, created a high-minded belief in Britain that GM food must be prevented from entering the country by any means necessary. For some people, this has meant that the end justifies the means.

This sentiment in Europe, and among some radical American groups, is by now beyond evidence and has become a matter of ideological passion. Some people also think that what these groups really oppose is not genetically modified crops, but capitalism and globalism.

Opponents in Europe have trampled and burned many empirical field trials of GM crops. This vandalism has been more than a few isolated events. Between the United Kingdom and the Continent, out of 150 to 200 sites where GM plants were being tested, eighty have been wholly or partly destroyed by anti-GM groups such as Greenpeace.[11]

In the most famous attack in England, Lord Melchett, Greenpeace's executive director, and twenty-six Greenpeace members at dawn on a day in July 2000 destroyed 2.4 hectares of Bt corn being grown at a government-owned farm in Dereham, Norfolk.[12] The sixteen men and ten women claimed that they had a legal right to destroy the crops to protect the environment. Greenpeace members had decided on ideological grounds that the experiment, designed to test whether or not the corn presented any danger to the environment, could not proceed. Arrested and tried, Lord Melchett and his merry green band welcomed the ensuing publicity, which (of course) they had carefully orchestrated. Despite overwhelming evidence that they had broken the law, they were acquitted.

Ironically, in 2000 officials announced that thirty thousand acres of GM crops had been accidentally sown in England and Europe on up to six hundred farms.[13] Rapeseed grown for oil used in margarine, ice cream, and chocolate had been mixed in with conventional seeds at a rate of one to one hundred. The revelation came only a day after Prince Charles's attack on the GM crops (see below). Moreover, products from the crops had already entered the feed of cattle and human food products. Despite the best efforts of Greenpeace to stop it, a field trial of GM crops already had been accomplished, and with no harmful effects.

In a related action, Greenpeace boarded a ship owned by Cargill Foods carrying soy with added genes and destined to be animal feed. "This ship is carrying a cargo that nobody wants and most people would like to see sent home," said a spokesman for Greenpeace.[14]

In late May 2000, Italian Green organizations disrupted a biotechnology conference in Genoa, vowing to create a mini-Seattle in front of five thousand riot police. Many Italian politicians who had initially welcomed the conference turned tail and embraced the popular, antitechnology mood.

In England in 2000, many environmental organizations advocated a "Five Year Freeze" on GM foods and GM crops. Prime Minister Tony Blair initially supported GM foods but, like the Italian politicians and sensing that the vast majority of people were against him, abruptly switched his position.

THE PRINCE OF WALES DEFENDS NATURALISM

During 1999 and 2000, Charles, the Prince of Wales, denounced the introduction of GM food to Britain. On May 25, 2000, he broadened his attacks to include use of genetically modified plants to alter the environment. Charles said, "Nature has come to be regarded as a system that can be engineered for our own convenience . . . and in which anything that happens can be fixed by technology and human ingenuity. We need to rediscover a reverence for the natural world . . . to become more aware of the relationship between God, man, and creation."[15] Charles claimed that science "lacks a spiritual dimension and should be used to understand how nature works, not to change it."

English critics emphasized Charles's lack of scientific training and the fact that his own organic garden in Highgrove only produced because he paid a dozen people there to work full time. Oxford biologist Richard Dawkins chastised Charles for having a romantic idea of the naturalness of traditional or organic agriculture, saying "I forgot who it was who remarked: 'Of course we must be open-minded, but not so open-minded that our brains drop out.'" Professor Hugh Pennington of Aberdeen University blasted Charles's call for a balance in nature: "There is no natural balance in nature. If science had not affected that balance, we would all be living 30 years less." Pennington said he would feel safer eating genetically modified food than produce from Charles's organic farm.[16]

Charles's talk embodied what many English thought: that society should be cautious in dealing with GM plants, both as crops and as food, and should not make potentially irreversible changes in environment or human food until complete safety was assured. Invoking God and the natural, Charles spoke for many people who had been shaken by mad cow disease.

ORGANIC FOODS POPULAR IN BRITAIN, BOVE A HERO IN FRANCE

As a result of their experience with mad cow disease and vCJD, the British in 2000 bought five times more organic food than in 1996, the year before mad cow disease and GM foods became known.[17] Whereas Americans in search of organic foods must search out specialty food shops that often also sell herbs and alternative medicines, the largest grocery chains in Britain all sell organic foods, often with huge letters proclaiming it "FREE OF GENETICALLY MODIFIED ORGANISMS."

Defenders of Britain claim that their love of organic foods is more than hysteria about food safety. They claim it is the embracing of small specialty shops, a different relationship with nature, and the British love of walks in the countryside.

In France in July 2000, Joseph Bove, an activist posing as a French farmer, went on trial in Millau, France, with nine other farmers and union officials for using a tractor, pick axes, power saws, and spray paint a year earlier to destroy a half-built McDonald's. Officially, they protested the use of hamburger buns containing genetically enhanced flour.

Many previous protests had occurred against GM food and the Americanization of French life, but this time Bove refused to pay his bail and went to jail for three weeks. His role in the McDonald's destruction had come the previous November, just before the WTO meetings in Seattle, where he played a prominent role. Bove and his codefendants arrived at the trial in a cart being towed by a tractor, resembling victims of the guillotine.

By shrewd tactics, Bove made his trial a referendum on GM food, fast-food, and the influence of Americanized agribusiness on French life. Over fifteen thousand people showed up for his trial in a Woodstock-like atmosphere with free rock concerts, a farmer's market, and soapbox speeches. The gathering of all these people made Bove's trial a front-page story in the *New York Times* and in most newspapers around the world.[18]

Yet the press rarely mentioned the real cause of Bove's actions: the imposition of high taxes on French Roquefort cheese and *pâté de foie gras* by American trade officials in retaliation for the French refusal to import beef from American cattle raised on bovine growth hormones (BGH). Such taxes hurt French farmers badly, given the huge American market. Ironically, American beef was probably safer than ever before, since the hormone was genetically pure rBGH, not the previously used, cadaveric-derived BGH (more on this later). But science rarely triumphs over ideology and national protectionism.

THE SEATTLE RIOTS

During the first week of December 1999, hundreds of naturalists protested in the streets of Seattle in the largest organized protests in the United States since the demonstrations against the Vietnam War in the early 1970s. Over thirty thousand descended on Seattle to attack GM food and globalist business.[19] At this Woodstock of antiglobalization, a small group smashed store windows and fought with police, causing a state of emergency to be called by Washington's governor. Suddenly, the radical left seemed very much alive in America.

Few Americans had previously paid any attention to past demonstrations against GM food by French farmers, Greenpeace in England, or farm activists in India. (In truth, American mass media rarely report on such events.) For activists against GM food worldwide, their coming-out party was in Seattle. In the summer of 2001, similar riots occurred in Genoa, Italy, at the G8 summit meeting.

The event that attracted the protestors was a meeting of the WTO, an organization that few Americans knew about. Certainly no Seattle official expected the WTO meetings to cause such impassioned protests.

And passionate they were: all during that week, magazine covers and the front pages of newspapers everywhere showed riot police in helmets and gas masks battling bloodied protestors, who broke windows of Starbucks Coffee shops and McDonald's restaurants (these two chains were seen as symbols of creeping American global domination). José Bove, the French activist, told the crowd, "We don't want to eat any more of that kind of food. We got to throw it into the sea."[20]

The United States and 134 other nations had set up the WTO, a small, Geneva-based bureaucracy, in 1994 to referee global commerce.[21] While the WTO may not personify globalization, at least it gave foes of globalization an identifiable target.

What else did people protest in Seattle? In part GM food, but also a lot of specific things they didn't like about globalization. Some steelworkers and small American farmers waged a protectionist fight against international companies, which they accused of clear-cutting forests in Indonesia and in the Amazon and of using child labor and cheap female labor in Asia to make carpets and running shoes.

The Clinton administration, which had agreed in 1998 to host the WTO talks in Seattle, was surprised by the large number of passionate protestors, as were most Americans, the people of Seattle, and the Seattle police, while America's domestic intelligence organizations had not foreseen such protests would naturally grow out of previous controversies over growth hormones in milk. Globalist companies for the Left had become what Communists and Shi'ite terrorists had been for the Right.

Not every country at these meetings accepted globalist goals: small countries with trade barriers wanted to keep protectionist advantages; Europe and Japan, unable to compete with the large-scale efficiencies of American farmers, wanted to block American food imports and continue their massive subsidies to their native farmers.

Condemned by naturalists, American trade unionists, and egalitarians as antidemocratic, secretive, and unelected to make decisions that affected billions, the WTO reacted by becoming secretive. During the protests in Seattle and its later meeting in Montreal, it lacked even a spokesperson or PR professionals. Naturalists argued that environmental issues posed by introduction of GM crops could no longer be considered locally but had to be addressed globally. No WTO official publicly rebutted their claims.

Belatedly, pro-trade and pro-GM food arguments emerged. Arguments about trade protectionism, the enemy of globalists, mixed with debates about GM food. The Index of Economic Freedom, copublished by the Heritage Foundation and the *Wall Street Journal*, asserted that, "The more open a nation's economy, the better off its people."[22]

For globalists, free trade over the preceding five years had fueled one of the biggest economic expansions in American history. Because other nations were buying North American soybeans, apples, CD's, computers, cars, and machines, as well as paying to watch American movies and television shows and reading American books, the surplus from such trade was paying off the national debt, reducing unemployment to record lows, taking millions off welfare, making many new millionaires, and boosting the standard of living.[23] As a result, the Clinton administration wanted expanded free trade to create more markets for American exports, while simultaneously wanting to protect abroad such American intellectual products as copyrighted movies, songs, and books.

Globalists claimed that the new international economy had helped not just American companies but companies all over the planet: sixty thousand transnational companies such as Nokia (Finland) and Honda (Japan) now produce 25 percent of the world's output, according to a United Nations study.[24] Before globalization, the Philippines had been "the sick man in booming Asia," but its economy soared when half a dozen international companies invested heavily there, especially in fields such as accounting and high technology, creating hundreds of thousands of good jobs.[25]

Of course, Americans hypocritically asked poor countries to lower trade barriers while they simultaneously kept theirs up. Nothing is more subsidized in the United States than farmers who grow sugar, which sells for three times what it would if foreign sugar could be imported at market price.[26] Production of steel and peanuts is also heavily subsidized. Indeed, huge American farmers enjoy the best of two approaches: domestic socialism through huge federal price supports ($22 billion in 2000) and use of American bullying abroad to force their crops into foreign markets.[27]

At the WTO meetings in Seattle, President Clinton adopted the view that GM food was safe and he pressed European nations to accept it. He implied that trade protectionism and irrational hysteria were preventing European acceptance of American GM soybeans, GM corn, and GM cotton. He pressed European delegates to the WTO not to subject American farmers to "unreasonable delays and unfair discrimination based on suspicion unsupported by the latest scientific examination."[28]

Globalist Fareed Zakria argued that poor people would actually be hurt if nations gave in to the demands of the WTO protestors:

> In fact, if the demonstrators' demands were met, the effect would be to crush the hopes of much poorer Third World workers—the original "indigenous people." Citizens of developing countries have only one possible path out of the horrifying levels of poverty, malnutrition, and disease in which they live: economic growth. And every country in history that has raised its living standards—including the United States—has done so by hitching its wagon to the world economy.[29]

For such globalist thinkers, rising wealth broadly distributed in undeveloped countries would mean better medical care, cleaner air, healthier water, and better all-around living standards. Indeed, some delegates from poor countries claimed WTO protestors "cared more about turtles than the people of poor countries."[30]

THE MONTREAL COMPROMISE

In January 2000, a month after the Seattle riots, the United States faced stiff opposition in talks at meetings in Montreal to negotiate world trade in GM crops. These talks continued the 1992 Convention on Biological Diversity (in which the United States could not officially participate because its Senate had never ratified this convention's treaty).

Two very different kinds of issues arose there: (1) the safety of GM food and (2) the safety of GM crops introduced into the environment. The two issues were resolved in different ways.

European consumers, led by environmental groups, had pressed for labeling of imported GM food. In the United States, Consumers Union recommended labeling so people with allergies, e.g., to peanuts, could avoid GM food. Because of previous experience with labeling hormones in milk, American companies feared such labeling as the kiss of death. (Whether that is true remains to be seen: consider warnings on packages for cigarettes.)

The metaquestion of what standard could a country use to block import of GM food cut beneath both issues. The United States pushed for a standard that some scientific evidence of danger from GM food would need to be documented before a country such as France could block import. European countries countered with a much vaguer standard, *the precautionary principle*, which could be adopted even in the absence of real scientific evidence of harm.[31]

After five days and one all-night session of negotiations, a compromise was reached where the United States bowed to concerns of other countries and accepted a standard in between the above two.

A lot of money was at stake in the results because, in 1999 alone, the United States exported $50 billion worth of GM foods. The deal was in part seen as a good idea because it was considered better to get at least some agreement signed than to continue under uncertainty.

The precautionary principle has now become famous. It is a rallying point for naturalists and egalitarians against GM food. Especially in Europe, it grew out of their continuing experience with mad cow disease.

There is no doubt that mad cow disease, and other daily food scares created by British newspapers, caused Europe's acceptance of the precautionary principle. This principle allows a country to ban import of GM food based on concerns about *potential* dangers rather than actual, proven dangers.

This principle was affirmed by the Montreal accord about the environment. The basic ideas was that once an aggressive, exotic species is introduced into a yielding environment, it might be too late to recall it.

The reasoning here is that recalling a GM plant is not like recalling a bad batch of Coca-Cola or Tylenol. So the precautionary principle does not allow introduction of genetically modified plants, e.g., "Roundup Ready" corn or wheat (which tolerate heavy doses of herbicide), based on fears that the genes conferring resistance could spread in a new country to other, genetically similar crops, creating superweeds. Similarly, genetically modified bacteria could not be unleashed in a fragile environment from GM seed.[32]

On the other hand, GM food that would be consumed by people did not need to be segregated (as European delegates had wanted) when being exported in large shipments. But such shipments had to be labeled to indicate that GM food was mixed in with non-GM food. The final accord had the precautionary principle working in tandem with advance notice, so countries could segregate imported seeds and food in advance, or not accept the whole, mixed batch.

In essence, countries can ban import of GM food without full, hard scientific evidence of danger but at the same time, *some* such evidence is required before a ban can be instituted. Hence, the compromise puts a lot of weight on exactly what counts as "some" scientific evidence—a great concern given what has happened in England with mad cow disease.

What the United States lost on was labeling of GM food. The new system creates two systems of food distribution: one for GM food and one for non-GM food. (Labeling was delayed for two years in the United States, to commence in 2002.) Food distribution "will have two channels," said Willy de Greef of Novartis, a world leader in genetically enhanced seed, "one with GMOs, one for those who don't feel comfortable with it."[33] He predicted that the price of non-GM food would be higher than that of GM foods, although that remains to be seen. (In late summer of 2001, the Bush administration was pushing Europe to drop the labeling requirement.)

VANDALISM AGAINST GM CROPS IN NORTH AMERICA

The University of Maine at Orono is a land-grant university and, as such, has an agricultural unit. In one department there, research scientists tested whether pollen from Bt corn would spread to non-Bt corn. They planted a few acres of Bt corn and then planted non-Bt corn around it for hundreds of yards: one square inside a bigger square.

In a story that made national news, and even though the test was designed to test whether the BT corn was safe, anti-GM food activists in 1999 mowed down the BT corn in protest. Ironically, had the pollen spread to neighboring non-Bt corn, then the naturalists' claim of danger could have been proven.

Perhaps the protesters understood that pollination had already occurred weeks before and that corn pollen is heavy, and therefore doesn't travel very far. In any case, scientists were able to determine that about 1 percent of the corn in the immediate vicinity of the BT corn—the first few rows—had indeed been pollinated by Bt corn. However, at the far edges of the outer square, only 0.04 percent of the non-Bt corn had received Bt pollen.

In the United States, thirteen other field trials of GM crops were attacked in 1999. An American group called Reclaim the Seeds, after it had destroyed a field trial of sugar beets run by the University of California at Davis, claimed that "Modern agri-business and genetic mutilation is a capitalist machine that must be dismantled."[34] A similar American group, the Bolt Weevils, asserted, "Corporations give back to the people death."

On May 24, 2001, a spokesman for the Earth Liberation Front claimed responsibility for burning down a research lab at the University of Washington Center for Urban Horticulture, where genetically modified trees were being studied that could produce more pulp for paper.[35] The vandals also destroyed one hundred of the three hundred remaining plants of showy stickseed, an endangered plant being carefully nurtured at the lab.

END COMMENT

Food is never just food anymore. As we shall see in later chapters, it is about power in countries where people starve, and for international agribusiness it is about money and maximal profits. But in this chapter we explored another aspect: food as a cultural and political symbol.

Genetically modified food symbolizes the clash between American and European cultures. It symbolizes European resistance to encroaching American cultural hegemony over the planet. But while Europeans can do little about American military might, or space exploration, or finance, they know what they like to bring into their kitchens. They know how they like to eat and where. They know they don't want McCafés to replace the real ones, and they know they don't want the American model of fast food to become the norm in Europe.

As we shall see in chapter 4, Europe experienced an astonishing number of disasters in its food industry, as well as some environmental debacles, over the past decade. The worst of these disasters were mad cow disease and hoof-and-mouth disease, but there were also many other scares. Thus, when it was discovered that

American companies had secretly introduced genetically altered food into European diets, people again felt betrayed and shocked. Their traditional universe seemed out of kilter, and America seemed to be rocking it.

This partly explains why GM food caused such a political explosion in Europe. But deeper issues also explain why GM food has become so politicized. Our entire food system has massively changed over the past century and we are only just beginning to appreciate these changes. The average person knows little about them, and should know more, because important questions may be at stake about the kind of country, culture, and food we have.

European countries such as France and Holland massively subsidize their farmers, as does Japan. North America also massively subsidizes farmers, but to different ends. To make money, North American farmers must sell many products, such as potatoes, wheat, and corn, to the rest of the world, in part because they have low profit margins and make the most by selling a lot of food abroad rather than a small amount locally. But as we shall see, even this vast American food machine is in trouble, because most farmers, especially some of the very biggest, get huge farm welfare from the government, raising questions of just allocation of public resources.

For now, the Montreal compromise is something everyone can live with. If GM food must be labeled, so be it. Once consumers understand that GM food is safe, and once marketers or scientists give consumers an added reason to buy GM food at the same price or lower (better taste or nutrition, longer shelf life), GM food will be a success. But GM food should not be forced down the throats of American or European consumers; citizens should accept it voluntarily, as such and clearly labeled. Only in this way will GM food ultimately be depoliticized.

3

Four Perspectives on Food

At first I paid attention only to taste, storing away the knowledge that my father preferred salt to sugar and my mother had a sweet tooth. Later I also began to note how people ate, and where. My brother liked fancy food in fine surroundings, and Mom would eat anything so long as the location was exotic. I was slowly discovering that if you watched people as they ate, you could find out who they are.

—*Ruth Reichl,* Tender at the Bone: Growing Up at the Table

Whether they realize it or not, many people think about food from basic assumptions about human nature and how the world works. Four such perspectives figure prominently in current debates about policies about food: naturalism, scientific progressivism, egalitarianism, and libertarian globalism. Each is sketched below as to how they characteristically view certain topics, such as accidents, starvation, the alleged inevitability of globalization, and genetically modified (GM) food.

NATURALIST

Naturalists believe that human societies have strayed too far from traditional forms of growing food. Human sowing, weeding, and harvesting *justify* the bountiful harvest. Although some scientific improvements have helped agriculture, especially the new emphasis on ecology, most new artificial intrusions should be resisted, especially if they only benefit conglomerates, devastate the land, brutalize animals, and destroy natural taste.

International agribusiness distorts earth-friendly practices in farming, abuses livestock for efficiency, and pushes families off the farmland they inherited. Values other than large-scale efficiency should control production of food. Wendell Berry, the most famous American naturalist, claims that 25 million Americans left family farms between the 1940s and 1970s, and that figure is probably way too small.[1]

Naturalists believe that food should first and foremost be safe. They embrace organic farming and believe with E. F. Schumacher that "small is beautiful." They would rather pay more money for food that is organic, local, and from a family farm than pay for genetically modified food that is cheap, imported, and raised in a monoculture in a developing country by a global business only known by initials.

Naturalists prefer this kind of food because of the *values* that the system underneath it supports. Such values are multilayered and complex. Naturalists champion the aesthetics of low-tech, family-run farms compared to giant hog-processing-factories run by huge conglomerates in Iowa and North Carolina. Naturalists argue that small farms treat animals and people better, and hence are more moral. For naturalists, not everything should be about wresting the most profit from the land and from farm animals. Naturalists believe that sustainable, environmentally uplifting practices that produce natural foods are better for the locale, the country, and the planet.

Food naturalism at times stems from beliefs about how God wants humans to consume food. "Natural ways" of growing food may mean "intended by the Divine" or "approved by God" or "in harmony with his Nature." Industrial and GM food may be seen as ungodly, as "playing God," or as humans exhibiting hubris.

Naturalists not only cherish food safety but also environmental purity. Some of the most prominent may care less about food than about dangers to the environment from monocultures and from new varieties of genetically enhanced plants. In its environmental meaning, naturalism dictates "Leave earth as we inherited it."

Naturalists love organic farming and fantasize that one day it can be universal across the planet. Mae-Wan Ho, a leading food naturalist-egalitarian and retired lecturer in biology at London University, says, "Organic farmers are artists and poets. They have a certain relationship with the land, and the trees are poems the earth writes to the skies. They have a love affair with their land. Peruvian farmers adopt plants in their gardens as family members."[2]

The Naturalist's Creed

Nature contains an indwelling wisdom that should be respected. Bees know their job and, when left to do so unmolested by man, do it well. They pollinate our fields, give us honey, and propagate their own kind.

Now a few scientists think they're omniscient and want to cast aside Nature's wisdom for short-term gain by making artificial plants. This is typical scientific arrogance and a typical failure to see the dangers in the bigger picture.

What tests have these scientists passed that insures their wisdom? Did they take any course in ethics for their Ph.D. in plant genetics, much less one on wisdom? Do you find the meaning of Nature, or life itself, in molecular genetics?

Why unleash a vast array of genetically souped-up organisms on our fragile ecosystems? Why seek new, potentially dangerous life forms, especially when much of our planet's plant diversity remains uncharted and is under threat of extinction? Shouldn't we protect the treasures we have rather than seeking new, possibly false ones?

Let's keep this genetic Pandora's Box closed. Whether you believe that God, evolution, or both brought our fields and streams to this point in history, what we have now flourishes in rough harmony with humans. We are not wise enough to surpass the majesty of a homeostasis that has evolved over millions of years. If we tinker with its molecular essence, we risk losing it all. Lord, forgive us, if we do.

PROGRESSIVES

Progressives believe that scientific knowledge improves human life and that, as biotechnology, such knowledge should be immediately applied to help humans. They believe that food naturalists too often invoke nature to resist progressive change, often by raising cries of a slippery slope, or (following the usual portrayal in movies television) by imparting bad motives to scientists.

Progressives believe in agricultural efficiency, which they say has led to reliable, cheap food for people in advanced countries and which has saved millions from starving. As C. S. Prakash of Tuskegee University asserts, "despite the nonsense being spread by anti-biotech activists, this [food] technology can actually improve environmental conditions while helping to boost world food production."[3]

Progressives generally scorn superstition, theism, and the let's-wait-and-see attitude of precautionary principles. A founder of Greenpeace, Patrick Moore, quit it in disgust when this naturalist organization corrupted the debate in England about GM food. Moore said, "I believe we are entering an era now where pagan beliefs and junk science are influencing public policy."[4]

People have always been irrational and superstitious about new foods:

> In the nineteenth century, the tomato was known as the wolf's peach, and Europeans and Americans believed it was deadly poisonous. In 1820, New York forbade its consumption and only relented when Colonel Robert Johnston announced that he would eat an entire bag of them outside the courthouse in Salem, New Jersey. Two thousand people turned up to watch him die, while a band played a funeral march. But Johnston ate the lot and announced: "This luscious, scarlet apple will form the foundation of a great garden industry."[5]

Progressives emphasize that someone must pay the opportunity costs of inaction. Progressives embrace science, technology, math, and universal ideas, which they hope will break down parochial ethnic, national, and religious differences.

In particular, scientific biotechnology and recombinant genetics have created safer medicines and healthier foods, and helped people control fertility (as well as overcome infertility). Human insulin is no longer made from beef insulin; dairy cows no longer get growth hormones from cadavers of dead cows; hemophiliacs get purified clotting factors not contaminated by HIV. In the early 1970s, most people feared "test tube babies" (in vitro fertilization) because of ignorance and the alarmist rhetoric of people such as Jeremy Rifkin. But just as it would have been a mistake to ban in vitro fertilization then, so it would be a mistake to ban genetically enhanced foods now, which Rifkin now condemns.[6]

EGALITARIANS

Egalitarians believe that the primary problem facing the world is neither lack of scientific knowledge nor lack of food, but the unequal distribution of world resources. Where globalists embrace libertarianism and minimal government regulation, egalitarians want greater national and international regulation to smooth out the natural inequalities of fate. As Mae-Wan Ho, a leading egalitarian critic of such companies, says, "Science is not bad, but there is bad science. Genetic engineering is bad science working with big business for quick profit against the public good."[7]

For egalitarians, both national governments and the United Nations must be guided by principles of social justice to create a more fair allocation of world resources. The fact that North America can grow lots of food should not be seen as giving it the right to rip maximal profits from starving mouths, but rather giving it a duty to feed the world. With Harvard philosopher John Rawls, egalitarians think that the function of public policy should be to mitigate fate's natural inequalities, not to exacerbate them.[8]

Egalitarians do not believe that markets will naturally create wealth for everyone by creating a rising sea, or that genetically enhanced crops will solve world hunger. Instead, they believe that international food corporations will use these new tools to maximize profits by exploiting poor people in new ways.

Vandana Shiva is a leading critic of GM foods and biotechnology. This Indian physicist turned social critic says,

> Decentralization and local democratic control are political corollaries of the cultivation of diversity. Peace is also derived from conditions in which diverse species and communities have the freedom to self-organize and evolve according to their own needs, structures, and priorities.
>
> Globalization has undermined the conditions for self-rule, self-governance, and self-organization. It has established a violent order, both in terms of the coercive structures needed to maintain the order, and of the ecological and social disintegration and violence that are products of that order.[9]

For egalitarians, famine will only be banished when land and wealth are redistributed. For them, starvation reflects poverty and unequal rights, and not lack of food or lack of knowledge about how to grow more of it.

Satish Kumar, a former Jain monk, editor of *Resurgence* magazine, and founder of Schumacher College in Devon, England, says that the "Green Revolution" of the last two decades "has brought more poverty and problems than it has solutions. When you introduce capitally and chemically intensive agriculture to Third World countries, you deny employers access to labor that is available in abundance, and make them reliant on capital and chemicals, which are in short supply. It isn't wise."[10] Intensive agriculture only benefits big international companies, he adds, not local peoples.

Kumar disagrees most with progressives about the alleged neutrality of science: "Most science is at the service of technology, business, and governments, and is no longer a free agent. Most scientists therefore have vested interests." Because science works for who pays the bill, it doesn't help people who can't pay. To the horror of progressives, Kumar would just stop some areas of research: "genetics, nuclear research, and some research into mood-altering drugs that would allow healthy people to alter their emotions by popping a pill."

The Egalitarian Creed

They're at it again, those Masters of the Universe. Only this time instead of junk bonds and commodities, they're ruling over food, drugs, and the very molecules of life. A dozen international conglomerates bought patents on important genes, genetically modified food, GM animals, drugs, and pesticides. Soon the planet will witness only one huge, vertically integrated company. And when this company controls all of life, it will make trillions in profits as it makes the average person pay, and pay again, for what is rightfully his: the plants, genes, and animal life of the earth.

These new Masters have names such as Novartis, Con-Agra, Glaxo-Welcome, Cargill, and Astra-Zeneca. Their stockholders demand maximal profits, and in pursuit of that these Jolly Green Giants seek to tower over the world's biological resources.

Profits do not flow by doing good works. Con-Agra doesn't make money by feeding the starving, giving them free genetically modified food, or telling its researchers to concentrate on what will help the poor to live longer. No, what makes the most profit is helping existing agribusiness (and by that, we do not mean the nation's struggling mom and pop farmers with small plots) make more money with less work but more investment.

Enter genetically modified crops. By integrating new chemical pesticides (Roundup) with genetically modified plants (Roundup Ready soybeans, cotton, and potatoes) designed to maximize their effectiveness, agribusiness can use less of the expensive pesticides, grow more crops, and spend less on killing weeds, all of which creates more profits.

Some scientists would have you believe that everything is fine with these companies profiting on Nature and Life, but every scientist pushing GM crops is on the take from some biotechnology company; these pro-GM food scientists are shills for this new enterprise. University science professors own stock in, or are partners in, for-profit biotechnology companies, so can we expect criticisms of GM food from them?

If you want to understand an issue in bioethics, really understand it: follow the money trail. Where you see a scientist passionately defending GM crops, look at his sources of money, look at the free trips he takes to Europe or Costa Rica, look at the grants he gets, look at the reduced teaching and the fellowships for his graduate students, and you will see why he thinks as he does.

GLOBALISTS

Globalists believe that better economies, not moral passion, raise standards of living, and in the long run, more people are helped by free world trade, specialization, investment of capital, and economies of scale than by international aid.

According to globalists, only one way exists for poor countries to protect their environments, save for the future to capitalize their own industries, and educate their children: creating an expanding, democratic economy that gives everyone more hope—and plausible hope at that—than all the egalitarian or Marxist promises of instant, revolutionary change. For globalists, a rising sea raises all boats. According to them, moral earnestness doesn't protect the Arizona desert or the pristine beaches of the Gulf of Mexico; they are protected by a vested middle class and thousands of workers dependent on tourism, jobs in fishing, and retirees who want their habitat protected. Such people contribute money to candidates and influence elections in ways that get things done.

Globalists such as Thomas Sowell of the Hoover Institution see foreign aid as an egalitarian handout. Such food Band-Aids won't help in the long run, but only jump-start a sound economy will. Globalists think that transfers of cash or food lack sustainability because altruism is a scarce resource. In contrast, human desire for profit is unlimited and more reliable. Globalists and egalitarians differ fundamentally on food because they differ about human nature.

For globalists, just as everyone shouldn't try to grow his own food, so each region shouldn't. More efficient, better, cheaper food can be grown by some countries with rich natural resources, while other countries should specialize in, say, plastics or caviar. Just as everyone shouldn't make his own shoes, so every country shouldn't grow its own soybeans. Whether as individuals or as countries, and once a stable international marketplace flourishes, specialization benefits everyone.

The family farm, loved by naturalists, is an antique. Hunger will never be solved if every country tries to grow everything, especially on small, inefficient, family plots. Wendell Berry writes of a romantic world of family farms, and some of us fondly remember feeding sweet silage to cattle, helping to put hay into the barn, and shucking corn to feed kernels to the chickens, but such labor-intensive practices can easily be romanticized. In the open-air Museum of American Culture, in Virginia's Shenandoah Valley on Interstate 81 between Lexington and Staunton, visitors can see how eighteenth-century ancestors grew flax and painstakingly took threads out of these dried plants at night by candlelight. Then the thread was woven. These processes took a hundred hours to make one shirt.

For globalists, family farms today are like our ancestors struggling to grow and make such shirts. They just aren't efficient, although they may seem romantic now, when we look back with selective memories. We should preserve such farm museums to remind us of how hard our ancestors worked, but such museums should not be models for the future.

Egalitarians hate the way globalists conceive of biodiversity as a financial resource, as "commercial interests, whose profits are linked to utilizing global biodiversity as inputs for large-scale, homogeneous, centralized, and global production systems."[11] For egalitarian

Vandana Shiva, everything about genetic food is wrong, ethically, culturally, politically, and metaphysically:

> The monopolizing control of the molecular monoculture mind is most powerful through the rise of the new tool of genetic engineering. As Jack Kloppenburg has warned: "Though the capacity to move genetic material between species is a means for introducing additional variation, it is also a means for engineering uniformity across species."
>
> The production of transgenic species has been achieved through the crossing of species boundaries, which [previously and rightfully] have been nature's way of maintaining distinctiveness and diversity. While the ecological impact of crossing these boundaries has not yet been fully anticipated or assessed, a few predictions are possible. For example, breeding plants for herbicide resistance is one of the largest areas of investment in agricultural biotechnology. The aim is to concentrate market control of agriculture into the hands of a few corporations. At the same time, however, it introduces new pressures for uniformity since crops not resistant to these herbicides cannot be grown in fields contaminated by their excessive use.[12]

Globalists tend to be libertarians. They advocate minimal regulation of trade: just enough to keep it hygienic, efficient, safe, and flowing smoothly. With Robert Nozick, libertarians value liberty in personal life and freedom to control property they've already acquired.[13] They resent efforts by the state to redistribute their property to other people who don't have as much.

Egalitarians despise the globalist attitude that Nature is something to be turned into private property, and their antipathy on this point makes them allies of naturalists in fighting GM foods and biotechnology. To again quote Vandana Shiva:

> The conservation of biodiversity, at the most fundamental level, is the ethical recognition that other species and cultures have rights, that they do not merely derive value from economic exploitation by a few privileged humans. The patenting and ownership of life-forms

> is ethically a statement of the opposite belief. . . . Even as the world becomes more and more uncertain and unpredictable, technological and economic models are being based on a linear paradigm that assumes total certainty and control. . . . When a pig or cow is simply treated as bioreactor, for instance, to produce a certain kind of chemical for the pharmaceutical industry, it can be re-engineered and redesigned without any ethical constraint.[14]

In 2000, Italy's Green Party minister of agriculture, who opposes use of biotechnology in agriculture and favors labeling of GM crops, said, "We must resist the American principle whereby the consumer cannot tell if part of his meal comes from Patagonia and part from Canada."[15] His statement drives a stake through the heart of Italian agricultural globalists.

In the 1960s, futurist Marshall McLuhan became famous by analyzing how a popular medium such as a battery-powered radio could change a culture, say, when an indigenous Indian of South America carried such a device a thousand miles upriver and introduced rock-and-roll music into his hitherto sequestered tribe. Today, the wireless cell phone has done the same, as it has spread in countries such as Finland, changed adolescence in Japan, and made life less lonely on vast Wyoming farms.

But if the previous paradigm of a hot medium was a wireless device for voice communication, today's new one must be the genetically modified plant or seed. Like battery-powered radios, an enhanced seed offers superior growth at reduced costs, can travel around the planet with FedEx speed, and just as with the radio or cell phone, revolutionize the culture it enters.

Despite the objections of Luddites and antiquarians, globalization is both inevitable and good. The new international economy will be one vast, seamless, interconnected web of food.

PROGRESSIVES ON GENETIC MODIFICATION OF FOOD

Progressives hail new techniques of gene insertion not only as permissible but as far safer than traditional techniques. In the oldest of techniques, farmers attempted to crossbreed plants and animals to create newer, better organisms. The mule was created this way, as a cross between a horse and a donkey. So also were seedless grapes and fine roses.

This process was tedious and random, as the researcher had scant control over the mixing of the genes. It is said that Orville Redenbacher created thirty thousand varieties of popcorn before settling on the one he wanted. After the DNA revolutions of the 1960s and 1970s, researchers gained some control in transferring some genes from an organism, but even these Mendelian techniques transferred a variable number of genes in a hit-or-miss process that was highly inefficient. As Britain's progressive Nuffield Council on Bioethics concluded, "Conventional plant breeding is often a matter of combining two sets of about 25,000 genes. Entirely new species have been manufactured this way, such as Triticale, a synthetic hybrid between wheat and rye, which was the result of combining 50,000 largely untested genes, 25,000 from each species."[16]

Rather than randomly mixing fifty thousand genes and wondering what result will ensue, researchers can now insert a known gene and more accurately predict the trait expressed. Under the conventional system, even if researchers were lucky and a good trait was expressed, they did not know which of the fifty thousand genes (or which combination) were responsible for the trait, so that a similar trait might be added to another plant.

For progressives, the most important point is that conventional techniques of plant breeding allowed many other genes to be transferred with the desired genes, and that these other genes would have unpredictable effects when introduced into new environments. Rather than being more dangerous than conventional breeding, *targeted gene insertion is actually far safer and more environmentally pre-*

dictable because there is 99 percent more control over unwanted genes and their unpredictable effects.

Much more precision undergirds the new techniques that create genetically enhanced crops. Transferring a few well-characterized genes into fruits and vegetables allows scientists to know exactly which gene is being transferred, so they can study its expression in many environments. ("Expression" of a gene—a technical word—often varies in conjunction with other genes and with its environment, so a gene's expression is not a set, unfolding, linear template).

Similarly, animal cloning could allow farmers to reproduce the exact animal without sacrificing thousands of also-ran candidates created by random breeding. So with animal breeding, progressives can accept a premise of naturalists but reject the conclusion. In other words, they can accept the premise that the suffering of animals raised for human food is bad, but argue that because gene-modifying techniques reduce animal suffering, genetic modification is not bad but good.

Thus for progressives, both animal cloning and plant enhancement offer—for the first time in human history—much greater control over the precise characteristics of new organisms. It is precisely such control that worries critics. Naturalists worry that claims about human ability to control such organisms are exaggerated and that wild strains may ruin the ecological balance. Animal cloning is a good example, because where animals at first seemed normal there are now worries about latent irregularities in the expression of genes created in new ways. Egalitarians also worry that big companies will use such control against people.

TERMINATOR TECHNOLOGY

A defining moment for both naturalists and egalitarians occurred when Monsanto sought to sell farmers seeds of genetically altered crops that would produce plants with no seeds. The background of this key event is as follows.

Biotechnology corporations have always feared intellectual theft, and recent innovations in agricultural technology increased such fears. Policing such theft has always been a headache for seed companies. In the previous decade, with supercrops such as Roundup Ready soybeans, Monsanto contracted with farmers who promised not to plant the seeds produced by their crops and to buy a new load of seeds from Monsanto each year.[17] To enforce the contract, Monsanto had the right to inspect contracted fields anytime within three years.

These contracts were still difficult to enforce, and Monsanto brought several well-publicized suits against farmers whom they claim broke their contracts. But if there were a way to sterilize the second generation of seeds, the contracts would be insured and surprise inspections would be unnecessary.

The Delta and Pine Land Company (DPL) found a way to do this. Melvin Oliver of the Lubbock, Texas, labs of the U.S. Department of Agriculture (USDA), invented the technique. He describes it as "a way of self-policing the unauthorized use of American technology and similar to copyright protection." In March 1998, DPL won a patent (in conjunction with the USDA) for a crop-seed sterilization technique they described as a technology protection system. The rights were quickly sold to Monsanto. In practice, the result was a normal, mature crop harvest, but seeds were sterile.

DPL and Oliver's system depends on a promoter sequence called "late embryogenesis abundant" (LEA) that activates a toxin when the plant's seeds are maturing. The researchers inserted a DNA sequence for the LEA promoter so that the toxin would prevent germination. At the end of the growing season, the LEA promoter would switch on the DNA sequence.[18]

But critics hated the idea. In a stroke of semantic genius, a naturalist, egalitarian organization called the Rural Advancement Foundation International termed the breakthrough "terminator technology."

Alternative techniques of sterilizing seeds, dubbed "terminator II," include plants developed by Astra Zeneca that will not grow unless activated by a certain chemical (which the farmer would buy from Astra Zeneca). (This reminds some of the technique for cloning dinosaurs used in the movie *Jurassic Park.*) Transgenic plants also exist whose natural pest resistance cannot be activated by anything except specific chemical signals (purchased, of course, from a seed company).[19]

For farmers in developed nations, this argument over seed property rights is simply an ideological problem. But sterilized seeds for subsistence farmers in developing countries mean economic dependence each year on companies such as Monsanto. Unable to save and replant seeds, farmers will be forced to buy seeds every year from biotech companies, a change from traditional agricultural practices.

Of course, naturalists critiqued such changes. Farmers have saved seeds for centuries. It's their "age-old right," protested Camila Montecinos of the Center for Education and Technology in Santiago, Chile.[20] Her organization urges governments to outlaw the new technology. If farmers are unaware they are buying terminator seeds, she has claimed, whole growing seasons could go to waste.

Egalitarians emphasized that poor farmers don't have the money to buy new seeds every year. Is biotech going to put an end to agricultural development in countries such as Argentina and Mexico?

Not quite, replied progressives. They pointed out that the terminator genes would be used on agriculturally innovative crop seeds, which would either boost production or lower costs of production through built-in herbicide and pesticide-resistance. Either the increased profits would make up for the annual purchase of seeds or poor farmers wouldn't buy them.

Monsanto claims that, as other herbicide companies have lowered their herbicide prices to compete with the Roundup Ready soybean system (seed and Roundup), the cost of seed and chemicals for growing soybeans in the United States has dropped by 15 percent since 1996. An industry group, the American Crop Protection Association (ACPN) claims that American farmers spent $167 million less (a 9 percent decrease) on soybean herbicides in 1998 than they did in 1997, as Roundup Ready soybean planting grew from about 9 million acres to 27.5 million acres.[21]

Voicing sympathy with globalism, DPL vice president Harry Collins even claimed that farmers in developing countries would not be ripped off by terminator technology, but that in fact such technology would be good for them because "it will help them become more production-oriented rather than remaining subsistence farmers."[22] For egalitarians, this is like a Carnegie, Rockefeller, or Vanderbilt saying that low pay is good for workers because they would squander more money if they were paid higher wages.

Egalitarians and naturalists also fear monopolization of the market for seeds. In 1999, DPL controlled over 70 percent of the cotton seed market, while an oligopoly of five companies controlled 70 percent of the seed corn market.[23] If bean, cotton, and corn farmers become completely dependent on biotech companies for new seeds, prices could rise unreasonably, because such companies could always adopt a "take it or leave it" approach. However, USDA spokesman Willard Phelps says that the government's aim is to "ensure the profits that are made from [the biotech companies' seeds'] introduction are not excessive."[24] Progressives reply that terminator genes would only be used on supercrops (such as Monsanto's Roundup Ready line) and that farmers will thus always have a choice to plant seeds of traditional crops.

What happened next was a stupendous victory for egalitarians and naturalists. The idea of being able to store seeds for next year's planting was a very ancient one and most people, including leading intellectuals, had strong intuitions that making poor farmers pay each year for seeds was "just wrong." Many people wrote and said so, and spokespersons for poor farmers, such as India's Vandana Shiva and England's Mae-Wan Ho, quickly became internationally famous for their unrelenting criticisms of terminator plants.

Before field trials of crops with terminator genes could be performed, lawsuits were filed by egalitarian groups against the USDA, charging it with failing to protect the public interest. In the face of mounting criticisms, Monsanto backtracked and reassured the public that such a technology would probably not be available until 2003. The company quickly announced that it understood the opposition's criticisms, especially that of developing country farmers, and that the technology would not be introduced until the issue of these farmers and all other consumer concerns had been considered.

This was not an overwhelming surprise. Although it had successfully conducted its stealth introduction of GM foods into the U.S. market, Monsanto perceived from the massive public outcry in Europe and in many other nations that there was a possibility American consumers would resist GM food. "We know," they announced, "that the success of biotechnology depends on its acceptance by farmers and the broader public."[25]

Although it is certainly a good change for biotech companies to pay attention to public opinion, the concession of Monsanto had some huge downsides. As we shall see later, terminator genes could protect the environment against outcrossing or "escaping" genes. The idea of terminator genes can do wonders against seed piracy and protection of intellectual property, and that protection is important because commercial science is bankrolled by private individuals hoping to see a return on their investment. However sad it is that so much scientific advancement is financially motivated, it is fact, and further advancement may cease if investors cannot be reassured that their investments will return a profit and if companies cannot protect their research from piracy.

NATURALISTS AND FEAR OF ACCIDENTS

The defining paradigm of genetic modifications of plants for naturalists is fear of unintended accidents. Naturalists ally with egalitarians over fears that large companies will cut corners and inevitably produce disasters. In recent memory, they point to the spill of oil of the Exxon Valdez in Alaska, the discovery of chemicals dumped at Love Canal, the near meltdown of the nuclear reactor at Chernobyl, and the explosion of the Union Carbide plant at Bhopal, India.

If we switch from fact to fiction, there are countless novels and movies about such accidental disasters. Michael Crichton's *Andromeda Strain* was about a crystalline life structure that threatened the human race, after being introduced to the planet by the reentry of a satellite. *Jurassic Park* was about the mayhem introduced on a small, jungle island by the reintroduction of dinosaurs cloned from the DNA of long-dead ancestors. Mary Shelley's *Frankenstein* is about a bad-looking but noble and warm-hearted artificial human who an uncaring scientist created out of arrogance. Kathleen Ann Goonan's *Queen City Jazz* (1994) marries nanotechnology to biology to create an ersatz future planet where no one starves or works, but where nothing tastes natural and things ultimately go bad because of a secret flaw that planners overlooked in their haste to solve the world's problems.

The media scare people about accidents because most people who write for popular magazines reflect what most people fear. Consider this quote about the promise of the new genetics from a 1984 piece in *Washingtonian* magazine, a periodical sold to affluent, educated, sophisticated readers:

> The outdoor spraying of gene-engineered bacteria that prevents food crops from freezing could, if the bug found its way into the upper atmosphere, play havoc with the world's weather by inhibiting the formation of snow. Similarly, gene-engineered organisms intended to clean up oil spills could themselves become an uncontrolled environmental scourge.[26]

The reporting here of course merely retells the ancient fable of Pandora's box, one of the most enduring stories of the Western world. Indeed, the power of the Accident Story helps explain why Europeans reject GM food.

Biologists themselves have created worries about unintended Big Accidents. Many species of plant have been intentionally introduced into a local environment with unforeseen consequences. The Japanese kudzu vine, introduced into the American South for erosion control, spread "like wildfire" and now chokes mature trees in many places. One could also cite Dutch elm disease, the Gypsy moth, starlings, Mediterranean fruit flies, and the mongoose.

One of the most famous biological accidents in North America is the introduction of the zebra mussel from the Caspian Sea region into the Great Lakes. The mussels choked intake pipes and encrusted navigation locks so severely that they could not open. Stripping algae from the water, the zebra mussels ingested PCB's and dioxin hitherto lying dormant in the mud at the bottom of these lakes, passing along the toxins to the birds and crayfish that ate the mussels, and hence up the food chain to trout, bald eagles, and even humans.[27] Naturalists think a similar accident could happen with GN crops.

THE FOUR PERSPECTIVES AND THINKING ABOUT FOOD

The four perspectives described in this chapter do help us think about food. The preceding discussion illustrated, for one, the egalitarian and naturalist skepticism about whether a for-maximal-profit food system can be trusted. In arguing this way, egalitarians especially provide *a reason* why the food system can't be trusted, that is, that the self-interested desire for profits by, for example, the owners of meat production that motivates them to cover up scandals about infected meat, spoiled meat, and problems in the system.

In demonstrating against biotechnology and GM crops outside a large biotechnology conference in Boston in 2000, protestors such as Sarah Seeds accused the companies of greed and indifference to the health of peoples of the world. "These are the people who gave us thalidomide babies," she said. "Now they want to give us genetically modified food."[28]

Naturalists come to the same conclusion from different premises: because an infatuation with scientific efficiency blinds meat producers to the wisdom of traditional ways of raising cattle, meat producers underestimate the true risks of their methods. This attitude is wonderfully captured in the naturalist phrase for genetically changed food crops: "Frankenfoods."

As we shall see in the next chapter, naturalists also site transmissible spongiform encephalopathies (TSEs), the human form of mad cow disease, as evidence of what happens when the food industry uses unnatural "cannibalistic" practices.

Interestingly, the globalist accepts the premise of naturalists about TSE, but rejects the conclusion. That is, globalists accept the premise that people are self-interested, but they believe that such self-interest is not limited to owners of meat production, but is shared by consumers and buyers. Enlightened self-interest spread globally shows that one safe standard will be in everyone's self-interest. Trade protectionism always fails in the long run, which explains why Germany's and France's initial reactions of banning English meat were convenient diversions when they should have been cleaning up their own abattoirs.

It is also true that some perspectives provide answers to certain questions but not others. The naturalist and egalitarian perspectives provide answers to the question of trust, but other questions are answered better by the other two perspectives. For example, let us take the question of the likely risks and benefits of eating meat.

Without going too much into the dangers of eating meat, which will be discussed in chapter 5, the real question here concerns what risks one is willing to take. For progressives, a small risk exists of *E. coli* infection from undercooked meat and an unknown, small risk of contracting a brain disease like mad cow disease also exists.

Given that they are unknown, how are we to evaluate such risks? Here there appears to be a fork in the road, with naturalists and progressives going in opposite directions. Naturalists want the scientist to *prove* that eating meat won't cause mad cow disease, whereas progressives demand strong scientific evidence that eating meat really causes it.

So we reach a standoff: the naturalist believes there may be a huge, silent epidemic of TSE-caused dementia in the population, caused by eating prion-infected meat, that will not surface for decades. The progressive believes that, absent any supporting evidence—such as might be obtained by taking biopsies of tonsils after tonsillectomies, which to date has showed no evidence of prions or TSE disease—claims of a silent epidemic are alarmist, irrational, and possibly, immoral.

In turn, naturalists and egalitarians believe that progressives put too much faith in science and its inevitable progress. Such meliorism, naturalists and egalitarians claim, ignores the fact that while theoretical science may be pure, its applications are not. One could borrow a phrase from Sartre and call this the scientific "problem of dirty hands": even if scientific knowledge is value free, what is done with it is not. If theoretical science is the apex of the pyramid, then life may be very different down at the bottom. Theoretical science may claim that meat *raised and killed correctly* is safe, just as genetically modified potatoes *grown and fertilized correctly* are safe. But whether these conditions exist is a matter not of science but of politics, finance, culture, agricultural ethics, and the law.

Another vantage point from which to contrast the four perspectives is motive. Naturalists and egalitarians try to capture the moral high ground and assign bad motives to applied scientists, agribusiness, and international corporations.

At the same time, naturalist organizations such as Greenpeace and egalitarian thinkers such as Jeremy Rifkin imply that only their motives are pure. But progressives won't cede them such high moral ground. As one defender of progressivism says, "If anything clearly emerges from this debate, it is that, when the veneer of pious rhetoric is stripped from the anti-GM food claims, their argument is simply one of selfishly seeking to impose their own fetishes and New Age beliefs on society whatever the costs to the rest of humanity may be. It is really questionable whether anyone benefits from opposition to GM technology (or from opposition to any genuine advance in science and technology) except those organizations that gain membership, funding, and above all, power from these Luddite actions."[29]

Although this is somewhat simplistic (me = good motive, them = bad motives), thinking about motives is sometimes revealing. In particular, egalitarians charge that genetically modified plants so far have only gone to help industrial farming in developed countries and have done nothing for the poor of the world. (Whether that is true or not will be discussed later.) Even if that is so, globalists retort, is it the job of a corporation with stockholders to feed the world? Isn't that the job of churches, philanthropy, and benevolent foreign assistance? If investors don't buy their stock and get a good return, biotechnology companies go out of business. Should not their motive be to make money?

END COMMENT

Progressives make a good case for the safety of GM food and for why we should back medical and scientific research. As we shall see in chapter 7, they also trump on GM food's role in ending starvation.

Naturalists have a strong argument who want to keep family farms and a countryside that looks like it is being worked by active farms. I spent summers as a child on my grandparents' farm in the Shenandoah Valley, a farm today that has been subdivided and subdivided, and on which at one time a developer contemplated putting trailers on lots along the Shenandoah River. National policy, state policy, and even counties must resist letting everything become suburban sprawl and, even if it makes food cost more, we should preserve traditional farming and farmlands for aesthetic and spiritual inspiration.

Egalitarians have made the world better by arguing that society and government should mitigate, not exacerbate, the natural inequalities of fate. When it comes to food policy, egalitarians give us a reason why mad cow disease went unreported so long: there was no money in preventing, reporting, or curing it. This should give us pause about whether ag science will tell us the whole truth and nothing but the truth when it comes to the safety of GM food. When in doubt, follow the money trail.

I think it is possible to combine the best of all views on a case-by-case, issue-by-issue basis. Good science may help us preserve the best of the natural. Cloning may help us keep alive species that are endangered.

Good science can work for a better, more just world. One can favor a worldwide movement toward greater equity of life prospects without being antiscience. Moreover, one has to distinguish between science and scientists, on the one hand, and those who use science to make money, on the other.

In any case, it is false that each of us must choose to be guided by just one simplistic frame. Most issues in ethics are complex, so simplistic worldviews rarely do justice to the interests of everyone involved. While these frames can contribute insights, they are not the whole story.

4

Europe and Mad Cow Disease

> Human beings have always found it easy to believe that wickedness might have physical consequences, and conversely that visible and tangible things "out of place" can reflect something evil being done. . . . Modern people, who pride themselves on escaping such a primitive concept as a perceived connectedness between morality and physical symptoms, have found a new meaning for the word "pollution": black ooze covers golden sand beaches, the sky rains poison, plants and animals grow stunted and die—and we conclude that the heedless human greed has come to haunt us in physical form. . . . Pollution has always meant matter out of place, and rules broken. . . . Eating food, cooking it, serving it, sharing it out, and passing it to others requires intensely intimate contact, both with the food and with the dinner companions. Pollution rules hedge food about, therefore, with particular fierceness.
>
> —*Margaret Visser,* The Rituals of Dinner: The Origins, Evolutions, Eccentricities, and Meaning of Table Manners

Many people who have faith in science and biotechnology believe in the safety of our food system. What happened in England and Europe during the 1990s tested that belief. During that time, Europeans began to fear the meat on their tables because people started dying who had naively trusted farmers and their public health officials. Within a decade, Europeans and especially the English went from a gullible trust in their food system to a deep cynicism.

Naturalists use Europe's experience with food as a template for thinking about whether agribusiness can be trusted about GM food. Based on what happened in Europe with mad cow disease, naturalists ask: If GM foods were discovered to pose dangers, would North Americans be promptly notified?

Beyond this question of prompt notification, we must understand how mad cow disease, and more recently hoof-and-mouth disease, caused Europeans to distrust everything about their industrial system of food. Regardless of the scientific merits of genetically enhanced food, can citizens trust governments that are in bed with international agribusiness to protect them from the dangers of new kinds of food?

Although this tale could begin in any recent year and with numerous examples, we begin in the Mediterranean Sea with purple algae, *Caulerpa taxifolia*, a.k.a. "killer algae."[1] Used in tropical fish tanks, the algae probably escaped in 1988 from an aquarium in Monaco. Supposedly, it couldn't survive the Mediterranean's winters, but like kudzu in the American South, killer algae proved the experts wrong, spreading from the shores of Spain to those of Croatia.

Killer algae symbolizes why Europeans don't trust European Union (EU) officials to protect them. In the early 1990s, EU officials debated where the algae came from, whether it would spread, and who was responsible for preventing its spread. Because of unclear lines of authority across national boundaries, no one took action, and the algae spread. Trusted officials let the public down.

This is not a rare story. Repeatedly urged to trust authority, Europeans and Americans have been falsely assured a lot by appointed medical and scientific officials over the past two decades.

From 1981 to 1985, American Health and Human Services secretary Margaret Heckler and American Red Cross president Joseph Bove (not to be confused with the French activist) both repeatedly denied that Americans (and especially hemophiliacs) had any chance of contracting HIV from contaminated blood during surgery. During those years, donated blood could have been tested for hepatitis B, but—in a risk-benefit analysis that considered the risk of not testing small—such testing was deemed too expensive. (Eliminating blood testing positive for hepatitis B would have eliminated perhaps 80 percent of HIV-infected blood, because affected donors often had both conditions.)

For thousands of Americans who received HIV-infected blood, these false assurances were death sentences. As many as twelve thousand Americans were infected with HIV and later died of AIDS; similarly, 60 percent of America's twenty thousand hemophiliacs became infected (from clotting factor pooled from many sources), and about a third died.

In the late 1980s, the English learned that imported American blood had infected twelve thousand English hemophiliacs, half of whom later died of AIDS. In 1985, French officials delayed using the American ELISA test to detect HIV-infected blood for the political objective of supporting a new, commercial French test for HIV. As a result of this foolish chauvinism, forty-four hundred French people became HIV infected, many of them hemophiliacs, and 40 percent of these died. In the late 1990s, four high French public health officials stood trial for this crime. Eventually, Health Minister Edmond Herve was found guilty, and state prosecutors charged "an immense breakdown of French medicine."[2]

While the French and English were trying to determine responsibility for these deaths from AIDS, a new scandal about "mad cow disease" erupted in England. Where during the AIDS scandal only hemophiliacs and hospital patients undergoing surgery were at risk, mad cow disease seemed to threaten anyone who had eaten beef in England within the previous decade. True to course, and as we shall see, a British health official went on television very early to assure the English and French of the safety of English beef, doing so by feeding his own daughter a hamburger from English meat on live television.

SPONGIFORM ENCEPHALOPATHIES

The discovery in Britain in 1984 of bovine spongiform encephalopathy (BSE) in thousands of British cattle is a story still unfolding in Europe. BSE is a species of the general class of transmissible spongiform encephalopathies (TSEs), which are only just beginning to be understood and studied, and for any of which no cure exists. The events behind the discovery of TSEs and how they can be transmitted in food directly led to Europe's subsequent skepticism about GM food.

And North America is not immune. In late 2000 a form of BSE was found in herds of deer and elk in the American Northwest and hunters were warned not to eat meat from these animals—a warning that American hunters generally disregarded. But how did these deer and elk get infected in the first place? No one really knows.

Cows infected with BSE eventually become very agitated, confused, spastic, and fearful, such that infected herds seem "mad" or crazy (hence "mad cow disease"). BSE is a transmissible spongiform encephalopathy (TSE). As it causes a similar, fatal degeneration of the central nervous system in sheep, we now think TSE in sheep is scrapie, a disease that has been around for hundreds of years in Europe. All of these encephalopathies produce small holes in brain tissue (hence "spongiform") and usually take years to wreak their damage.

Hans Gerhard Creutzfeldt first published a report on what is now called CJD in 1920, and his report was followed by another by Alfons Jakob in 1921. Creutzfeldt-Jakob disease (CJD) was named for these men. In 1976, Carleton Gajdusek and Baruch Blumberg were awarded a Nobel Prize for proving that CJD can be transmitted via brain tissue from one person to another.

In 1972, a patient of American scientist Stanley Pruisner died of a dementia that Pruisner identified as CJD. Pruisner thought that scrapie was a good model for CJD and—along with a few other scientists around the world—began to study it. To Pruisner, CJD seemed to make the human brain degenerate like scrapie in sheep and like BSE in cows, although with a longer incubation. No cures existed for any of these diseases.

At the time, it was thought that a species barrier could not be breached, such that a TSE could not be transmitted from one species to another, even by blood-to-blood injections. A fortiori, scientists thought this species barrier existed when meat from one species was eaten by another, especially if the meat was cooked before being eaten.

THE DISCOVERY THAT ENCEPHALOPATHIES HAD BEEN TRANSMITTED

In 1956, infertile women first received hormones derived from pituitary glands taken from the brains of human cadavers. Between then and 1985, very short children in England and America who lacked human growth hormones (HGH) received cadaver-derived injections of HGH.

In 1985, Alison Ley died from what was determined by autopsy to be CJD. She had received HGH as a child. In the same year, three American children who had received HGH also died of CJD, but their deaths received little publicity in the United States. Later that year, the British Health Department warned that injections of HGH could transmit CJD. Injection of short children with HGH was immediately stopped, but by this time over two thousand British children had been injected.

Although the Nobel Prize was given in 1976 for the discovery that CJD could be transmitted by injection of one human with brain tissue from another, it was not until nearly a decade later, in 1985, that British medical authorities ceased the practice of injection of cadaver-derived pituitary hormones to short children.

Pruisner suspected that cadaver-derived hormones had transmitted an encephalopathy to these children and caused their deaths. It would take him many years to gather evidence for this hypothesis.

In 1989, four Australian women, who years before had received cadaveric pituitary hormones during their attempts at assisted reproduction, died from CJD. Widely used in in vitro fertilization in

America, England, and Australia, this hormone may have infected other women. Since the incubation of CJD takes many decades, and because there is no treatment, these women were not notified of their possible infection. If they had been notified, it was determined, some would have suffered psychological distress over their possible infection, and some would have experienced suggestible symptoms. To this day, some women who attempted or used in vitro fertilization in the 1970s and 1980s may not know of their exposure to hormones that might have contained CJD.

In 1999, English orthopedic surgeon Neil Kreibich's family received £1.4 million in damages because he had been treated as a child with infected HGH. Also, thirty-six people, after taking HGH as children, were awarded from £5,000 to £340,000 for crippling psychiatric illnesses brought on by fear of having been infected with CJD. On February 17, 2000, Rachel Gwilliam, who had been injected with HGH as a child, died in England at age thirty-three.

By 2000, over thirty-two British children with pituitary dwarfism had died of CJD. Their deaths, like those of young adults in Britain from vCJD and of American Kimberly Bergalis, whose dentist had infected her with HIV, were often accompanied by angry accusations from the victims' families.

MAD COW DISEASE

In 1982, before the first cow died in England from BSE in January 1985, Stanley Pruisner concluded that something entirely new was transmitting this disease, what he called a "proteinaceous infectious particle," or "prion." His heretical hunch would prove to be perspicacious.

A smug medical establishment disagreed. No one had ever heard of a "prion" or had any idea what it did. No one believed that a protein could transmit disease, especially in meat that had been thoroughly cooked. All known modes of transmission of disease could be prevented by heat, irradiation, chemicals, or ultraviolet light. Yet if Pruisner was correct, prions escaped these processes, so that humans could be infected by cooked meat. Astonishingly, if prions existed, this hitherto-unknown way of transmitting disease to humans would be radically different from viruses and bacteria. Scary stuff.

A summary in 1998 in *NeuroNews* explains why Pruisner's views met so much resistance:

> [Pruisner] was ridiculed at first, since this went against all that was known at the time; that all infectious agents had nucleic acids. They were bacteria or viruses, that carried genetic information in their DNA or RNA. What Pruisner was suggesting was that phenotypic information could be carried by protein alone, from protein to protein, and from one animal to another. He concluded that the protein must be the agent, in part, because the infectious agent survived treatments that kill all other known infectious agents (those with DNA or RNA) and the protein he isolated was especially resistant to damage from heat or radiation. . . .
>
> The DNA sequence was determined for the prion protein. This led to the discovery that all mammals had the DNA that made prions, and they were everywhere. Through further research, it was theorized that prions can assume an alternate structure and that once a prion molecule is in this alternate conformation, it can catalyze the same conformational switch in other prion proteins. Thus, the aberrant prions, which are resistant to protease, solvents, and high temperature (while the normal form is not) accumulate in the cell and lead to the neurodegeneration observed in prion diseases. This explained why material from a diseased animal could cause a disease in another animal. The aberrant prion was causing the native prion protein to change conformation.[3]

The relevance of this tale later became clear to the British: food authorities resisted information about new kinds of dangers and sided with the food industry. If adding genes from animal species to plants caused similar dangers—for example, genes from cold-water fish transmitted to potatoes to create resistance to frost—would food authorities also be slow to react? As Hume said, a wise person proportions his belief to the evidence, and Europeans already had a list that included killer algae and HIV-infections in blood and pooled-blood products; now, they added meat infected with transmissible encephalopathies.

But we are getting ahead of our story, for first we must understand the colossal failure of trust by the beef industry that created this epidemic in the first place.

HOW BRITISH CATTLE GOT INFECTED

We know that, starting in the early 1980s, bovine growth hormones (BGH) taken from the pituitary glands of brains of slaughtered cattle were injected into cows to increase growth. One of the original, slaughtered cows was undoubtedly infected with BSE. Hormones from this cow were pooled and injected as BGH into many British cattle, spreading BSE.

Anne Maddocks, a retired senior medical scientist in infectious diseases at St. Mary's Hospital in London, repeatedly claimed that, "The promiscuous use of pituitary hormones in cattle led to BSE in the same way that they led to CJD in humans. The timing of the deaths in cattle and humans who were exposed to pituitary hormones is very compelling."[4]

According to the final *BSE Report*, BSE originally arose from a spontaneous mutation in a single cow in the early 1970s.[5] However, this conclusion is still contended, as some scientists still believe that scrapie crossed the species barrier from sheep to cattle, probably by cows eating infected MBM (meat and bone meal). (The emergence of a TSE in elk and deer in Montana may support the hypothesis of spontaneous mutation.)

In early 1985, a cow later dubbed "Cow 133" died of a mysterious illness. By 1986, BSE was officially accepted as a new kind of death in cows. In 1987, the veterinary office of the British government admitted that BSE had infected some cattle. The following year, the government prohibited ground animal remains from being added to feed for cattle and sheep.

Besides direct injections of hormones into cattle, British livestock could have been infected by what they ate. How was that possible? The idyllic picture of happily grazing cows eating summer grass or winter hay is not the picture of modern agribusiness. As the science editor of an English newspaper wrote,

> the diet of cows is an unpalatable story everywhere. In the last few decades, cattle in Britain, Europe and the US have eaten human excrement, chicken manure, and even chickens. They have, notoriously, eaten other dead cows. They have licked sump oil, chewed cement, and munched down plastic chippings as re-usable roughage. . . .
>
> In a business in which input costs have to be low and outputs high, grass and hay are no longer enough: for the past century, agribusiness has been looking for cheap proteins, fibers and fats to beef up the output from the cattlesheds.
>
> Some of it grows to hand: sheep and even cattle will eat seaweed, and browse from foliage. They will also consume oat and barley straw, vegetables and windfallen fruit. But to produce high levels of protein in milk and cheese, or in meat, they need a high protein, high energy diet.
>
> The ideal is to use protein and fats that would otherwise go to waste. Some of the waste was right there in the barn. So British farmers experimented with human and cattle excrement—heat-treated—two decades ago. This was possible because neither human nor cattle digestion is very efficient, so there is still a lot of nourishment even in manure.
>
> There were other farm wastes to be recycled. Farmers struggled with the cost of disposing of dead animals. But even dead, these animals were nutritious. So they began feeding cattle with chopped up, heat-treated animal carcasses as a source of cheap crunchy protein at the beginning of the century.
>
> Both experiments came to a stop in the wake of the BSE epidemic.[6]

HOW DID BSE INFECT HUMANS?

It was already known in the early 1980s that a spongiform disease called Kuru, a prion disease normally transmitted by ritual cannibalism in New Guinea, could be transmitted from eating infected human flesh.

If a transmissible encephalopathy could be spread by medical treatments using growth hormones (which had been treated to prevent the spread of infectious diseases), could eating beef given similar hormones also spread diseases?

Specifically, although it was thought impossible at the time, could eating the meat of BSE-infected cattle infect humans? More specifically, when a new form of spongiform encephalitis, called "variant CJD" or vCJD, was identified in the mid-1990s, could this have come from eating BSE-infected meat? For unknown reasons then and now, this new form had a short incubation period and infected only young adults.

Prima facie, this mode of infection would be unlikely, since similar cases of CJD had been spread blood to blood (injections of growth hormone, transfusions, possibly sharing needles). Moreover, when cooked well, meat could not harbor known pathogens.

Now we know in hindsight that the answer was yes. According to the final *BSE Report* by the British government in late 2000, what misled scientists was an early conclusion by epidemiologist John Wilesmith, who concluded that BSE was a form of scrapie and had been produced by the rendering process for turning cattle remains into cattle feed.[7] Because it was known that eating scrapie-infected meat could not harm humans, Wilesmith concluded that eating BSE-infected meat could not harm humans. According to the *BSE Report*, that was his fatal error.

In America, and before recombinant DNA techniques created pure forms of hormones, bovine growth hormone (BGH) was obtained from the brains of dead cattle. BGH increased production of milk and weight in cows. Because it was hard to obtain a lot of BGH from cadavers, not much could be easily and cheaply given. In the latter 1980s, the recombinant form, rBGH, made production inexpensive, easy, and pure of infections. But just when it was safe to give it to cows, Jeremy Rifkin, exploiting people's fears of phrases such as "genetically engineered," sounded alarms as if the genetically created rBGH was much more dangerous.

So ironically, while Rifkin scared Americans about the alleged dangers of milk with genetically created hormones, use of such biotechnology *virtually eliminated* the dangers of infection of BSE. Rather than becoming more dangerous, such milk was actually much *safer*.

Similarly, when genetic engineering created pure forms of clotting factors for hemophiliacs, obviating the need for pooled sources, hemophiliacs who got the pure factors no longer risked acquiring HIV, hepatitis, and other blood-borne infections found in donated blood.

(Lest North Americans get too smug about not having mad cow disease, it should be stressed that meat producers in both countries do the cheapest thing to produce the most product. In America, the soybean capital of the world, feeding soybeans to cattle is cheaper than feeding them recycled animal products, although American farmers employed this "cannibalistic" practice too until it was banned in 1996. In England, importing soybeans is more expensive than feeding recycled, dead cattle. But if it were cheaper in North America to feed cattle the recycled remains of dead cattle than to feed them soybeans, it probably would have been done more. And if we had also had a baseline case of BSE in a cow whose remains were recycled into animal feed, we would also probably have had mad cow disease in our midst.)

Meanwhile, continental Europe in 1989 banned import of British cattle born before 1988, when new controls were instituted, citing the possibility of its *own* cattle becoming infected. In 1990, the European Community banned import of cattle born before June 1989. By July 1993, the one thousandth case of BSE had been discovered in British cattle.

The European Community had earlier banned import of all British beef to the Continent. This resulted in a famous false assurance in 1990 by the chief medical officer of Britain, John Gummer, who fed his daughter a British hamburger on national television. Other politicians asserted that eating British beef posed no danger to anyone. British beef was as safe as British blood, British chauvinists said, while some voices in the wilderness remembered a similar assurance by American health and human services secretary Margaret Heckler about the safety of blood in the early 1980s and later assurances by English officials about the safety of their own blood supply.

In 1995, Stephen Churchill, patient zero of vCJD, died at age nineteen. A decade after a veterinarian had first discovered BSE in British cattle, five years after John Gummer had fed a British hamburger to his daughter on television, this fatal brain-wasting disease killed its first victim. Instead of fifty or forty years, the time between infection and death was . . . what? Ten years? Twenty? No one knew, but certainly it was less than with traditional CJD.

That year saw two more cases of vCJD reported. At this point, the body count from vCJD stood at three and the body count from CJD from patients infected with HGH was also growing. In March 1996, medical authorities finally acknowledged that a new variant of CJD was killing young adults.

As more cases surfaced of vCJD, European media ran daily stories that an AIDS-like disease had infected British beef, which could now be infecting future victims. "Mad" cows staggered as if drunk, while emaciated young English men and women developed spastic muscles before dying.

In March 1997, 100,000 British cattle were killed. Britain and Europe banned most (but not all!) parts of slaughtered animals in feed for livestock. The United States followed suit. A comprehensive system began to trace cattle, should any later be identified as a vector of disease. The government also banned sale of "beef on the bone," e.g., T-bone steaks and beef cut for barbecued ribs. By 2000, over 176,000 British cattle with BSE had been identified and killed.

The deaths of such enormous numbers of cattle created huge, eighty-thousand-ton mounds of bovine waste at various sites in Great Britain. These mounds were slow to go away because the only incinerator capable of achieving the required one-thousand-degree temperature could handle only fifteen thousand tons a year. The government secretly began to store all this bovine waste in old airplane hangars from World War II.[8]

In 1990, British food authorities had denied that anyone could contract vCJD from eating British beef. In 1997, red faced, they recanted. By then, twenty-three of the United Kingdom's young had died of vCJD.

Some saw parallels here with the first years of AIDS because both diseases struck down young adults as their first victims; because a mysterious, previously unknown agent was infecting people; because the disease had an unknown incubation (possibly short in the young, possibly much longer in others); because people rejected the obvious candidates for spread of the disease (unprotected sex, eating beef); and because some bigots hinted in both cases that victims must have "done something unnatural" to acquire such horrible diseases. The history of medicine shows that, when confronted with new infectious killers, people show a distressing tendency to first blame the victims.

Public recantation of their prior assurances undermined the authority of British food officials and food scientists: they seemed to always support food industries, not protect British consumers. (Authorities finally showed they had learned their lesson when hoof-and-mouth disease appeared in 2000, and they took decisive if unpopular action to stop its spread.)

Some people say it is still uncertain whether vCJD has a five-year or fifteen-year or even (like regular CJD) thirty-five-year incubation. (Infection by CJD is most reliably confirmed by autopsy, but prions can be detected in tonsils by biopsy.) Suppose the young people who have died are merely the canaries in the coal mine for this disease, the ones with genetic hypersensitivity who die first. If so, thousands of Europeans could now carry a ticking time bomb inside. If the worse scenarios turned out to be correct (as they were about HIV infection in Africa and Thailand), things might be very bad indeed. Scientist John Collinge claimed that vCJD rates of "biblical proportions" were coming and that the disease would strike down Europeans, especially from now-dormant CJD that would surface decades later.[9]

It has been a mystery from the start why vCJD mainly kills young people. One would think that it would first kill the elderly, those with compromised immune systems, or those who had been voracious meat eaters. But that does not appear to be the case. In this regard, it resembles the great American flu epidemic of the early twentieth century, in which almost all the million victims were young. According to the best theory today, those who were older had been exposed to similar strains of flu over the previous decade, before the very potent virus appeared, and hence had built immunity. The youngest had never been exposed to any flu and, hence, had built up less immunity. Whether a similar explanation could be true for vCJD has not been investigated by anyone the author has ever read.

In 1998, the English High Court ruled that the government had been negligent in allowing infertile women to be treated with cadaver-derived HGH. European newspapers and television stations drew the lessons from BSE/CJD for GM crops: "GM crops, like BSE, are a scientific minefield. Once again, the Government is in disarray, and once again we hear whispers of well-connected interest groups getting their own way."

On August 26, 1999, the first British beef in three years was served on the European Continent. Germany and France, citing their higher standards of food safety, refused to allow its import, creating a crisis in the EU.

By October 1999, the body count from vCJD was up to forty-six. English grocers and consumers were urged to boycott French food in retaliation for the continuing refusal by the French to accept British beef, a refusal backed by most French citizens. Reeling from their scandals over AIDS-contaminated blood, the French were wary. Germany also rejected British beef.

In 1997, Mathew Middleton died at age nineteen, making him the seventeenth victim of vCJD in the United Kingdom. A teenager who could wolf down a half dozen hamburgers at a time and who worked as a chef, he first felt terrible aches in his legs, followed by slurred speech, hallucinations, and inability to walk without stumbling. His physicians first suspected illegal drugs, then early-onset Parkinson's disease or multiple sclerosis. When neurologists diagnosed vCJD, physicians told the family, "Don't tell the press."[10] After his diagnosis, Mathew required twenty-four-hour nursing care as his brain quickly atrophied; he died within a month.

The vCJD body count by June 2001 stood at over a hundred in the United Kingdom. Three others had died in France and one had died in Ireland. Because only autopsies of brains can positively identify CJD, because such autopsies are expensive, and because they were not done before 1999, the true number of deaths from CJD is almost certainly higher. Officials worried that a bigger incidence had been silently incubating for ten years and could break out.[11] By its end, British authorities said, vCJD could strike down a minimum of several hundred British citizens or as many as 136,000.[12]

OTHER PRODUCTS AND TSES

Meanwhile, what about humans infected with CJD or vCJD who donated blood? What if such blood was used to create pooled-plasma products, collected from many donors? Had the lessons of AIDS been learned in England?

British medical authorities had previously been slow to act against dangers to the public's health. In February 1989, almost a year after BSE had been identified in cattle, medical officials allowed a half million liters of medicine that had been made with cadaverous bovine material to continue to be used in skin tests and vaccines. People undoubtedly were infected who, had authorities acted quickly and decisively, could have been spared.

In November 1999, British women received a jolt when they learned that some cosmetics they used during the 1990s could have been infected with CJD. Antiaging creams made from a cow's spleen, placenta, or thymus could carry BSE, and hence might transmit CJD. The EU banned such sources in cosmetics in 1997.

Then another shock hit Europeans: people could have been infected with vCJD by having received vaccines for measles; a combined vaccine for whooping cough, diphtheria, and tetanus; or a skin test for tuberculosis. All three products used sera derived from British cattle.[13] One BSE-infected cattle could have contaminated all these vaccines and tests.

Throughout this crisis, the status of science in Britain, or perhaps, more accurately, the status of government policies based on science, took a beating. "Policy in relation to BSE was not fundamentally based on science but on a lack of it," concluded Lord Phillips, the chair of the official British inquiry into BSE.[14]

Meanwhile, an ethical question arose over the possibility of a presymptomatic test for CJD. Such a test would screen out CJD-infected donors from blood and plasma, but like the test for HIV in the early years before the first treatments, who would want to be tested for CJD when no treatment existed? Do governments and politicians really want to know if vCJD is dormant and slowly growing in millions of people who don't know it? How many people were infected from eating meat? Blood transfusions? Sharing needles? Cosmetics? Assisted reproduction? Sexual intercourse? Are they, or is society, ready for what might be discovered from such testing? To provide the new, extra monies and resources?

In 1997, Stanley Pruisner received the Nobel Prize in Physiology or Medicine for his earlier discovery of prions. This was controversial because many researchers believed that prions were not actually infectious mechanisms but were either receptors for an infectious virus or vehicles that carried virinos, tiny viral particles. The Nobel committee defended giving Pruisner the prize, citing Europe's intense interest in TSEs and his discovery of a new method of infection (regardless of what prions turned out to be). The existence of prions is much less controversial in science today. In 2001, when Pruisner became vindicated and a hero, he gave the distinguished Shattuck Lecture (always published in the *New England Journal of Medicine*), in which he claimed that the study of prions offered "a unifying concept of degenerative brain diseases," including Alzheimer's, Parkinson's, and the various lethal dementias.[15]

OTHER FOOD AND ENVIRONMENTAL CRISES IN EUROPE PRIOR TO GENETICALLY MODIFIED ORGANISMS

In 1998, the Belgian food industry suffered a crisis when food fed to livestock became contaminated with the chemical dioxin, a by-product of chlorinated polycarbons and a chemical thought to be highly toxic to humans in several ways. Either motor oil or rancid vegetable oil at a rendering plant had caused the dioxin to enter chicken, pigs, and cattle raised for human consumption.[16] This contamination resulted in recall of food exported from Belgium to other European countries.

Worse, officials in Belgium later admitted that residents did not realize that wastes "from slaughterhouses, tainted with waste from toilets, showers and cleaning products, regularly ended up in Belgium's food chain."[17] Scientists from the toxicology unit of the University of Leuven wrote in a letter to *Nature* that toxic polychlorinated biphenyls (PCBs) had entered the food chain from animal feed contaminated with motor oil. Amazingly, Belgian "chickens and eggs had been contaminated with up to 250 times the normal tolerance level for PCBs—200 nanograms per gram of fat. Pig meat was less affected, with up to 75 times the tolerance level, while beef was effectively free of danger."[18] PCB's are believed to cause cancer in humans, although this claim is controversial.[19]

Understandably, clerks around the world immediately stripped Belgian foods from supermarket shelves, costing Belgium millions in sales. After these revelations, the average Belgian began to doubt the government's ability to protect the public's food. A similar story rocked France, where untreated sewage and sludge from septic tanks contaminated animal feed.

These events explain the great Coca-Cola scare in Belgium in the second week of June 1999, when forty-three schoolgirls mysteriously became ill after drinking Coke products. In its slow, reluctant reaction to this news, Coke followed the example of Perrier in a similar case in 1990, when the French company was slow to recall batches of its mineral water containing traces of benzene.

Coke tested its products and found no health problem, but other people across Belgium and then France reported illnesses. A week after the first reported illnesses, 101 people reported being ill. Some were admitted to hospitals for "hemolysis," or "excessive dissolution of red blood cells."[20]

As reactions spread across Europe, the next week 249 more people claimed to be sick, and Coke realized it had a big problem on its hands. Three weeks into the crisis, Coca-Cola's CEO flew from Atlanta to Europe, closed the bottling plants in Belgium, and announced that the cause of the reactions had been found: an inferior kind of carbon dioxide used to produce bubbles in Coke had inadvertently created a smell of rotten eggs. In addition, creosote on some wooden pallets had gotten on some cans. Evidently, the victims had smelled rotten eggs or creosote while drinking Cokes and concluded that they had ingested something rotten.

Although no microbial infection was ever discovered, Coca-Cola suffered a public relations disaster because Europeans perceived the problem as one of trust, not of evidence. In protecting them from dangers in foods and beverages, Europeans began distrusting both international, American-style companies as well as their own national food authorities.

On top of these scandals, Europe was rocked in February 2000 by revelations that cyanide had spilled from a gold mine in Romania and had killed most of the plant and fish life in two major rivers pouring into Hungary and Yugoslavia. The top environmental officer of the European Union called the spill "a major environmental accident" and, to the peoples living along the rivers, "a catastrophe."[21] A similar story later in the year revealed that a nuclear plant in Russia had leaked the highest levels of radiation ever measured—way beyond measurements of previous leaks—and destroyed all life in downstream rivers for hundreds of miles.

Suddenly, industries using scientific products in Europe seemed suspect, even evil. Europeans asked whether some sane controls could be instituted to keep new technologies from wreaking havoc on the environment.

TWO STEPS FORWARD, ONE STEP BACKWARD

In Europe, exactly what the danger is from eating meat remains controversial and partly a matter of cultural attitude. For example, Germans love sausage, some of which is made from brains (Kochmettwurst). Should German health authorities ban such sausage because some German cattle may be infected with BSE? Should the EU ban German sausage for export? Would Germans accept this?

This raises the interesting question of whether the German Green Party, eager to ban genetically modified plants, will push for banning knockwurst and other sausages in Germany and Austria. These meats could be infected with TSEs. Given the long incubation of TSEs and the problems of industrial livestock production, the dangers to the health of Germans are probably far greater from eating sausage than from eating enhanced food crops.

The year 2000 began with *Le Figaro*, perhaps the leading newspaper in France, announcing that banned protein and banned bone matter from slaughtered cattle had been found in feed given to French cows. Also, Switzerland put to death twelve cows infected with BSE. Problems with the safety of meat seemed to be spreading all over Europe.

The animal neurology unit at Bern University in Switzerland predicted that BSE would die out by 2007, although it had still not been proven that BSE could not be passed along to calves. The French believed it would not be eradicated in France before 2010.

The British government in 1996 banned use of slaughtered cows' offal, spines, brains, and bones in the feed of other cows. But because of pressure from the cattle industry to protect profits, and because prions had never been proven to be found in blood, the government did *not* ban other organic matter from slaughtered cows such as blood, tallow, and gelatin. Such practices continued despite the warnings of John Collinge, a professor of prion research at Britain's Medical Research Council, that "All cannibalistic recycling is potentially dangerous and I have said that repeatedly."[22] (Because of such concerns, America in August 1997 banned cattle feed containing dead sheep or cattle.)

Gelatin, made from crushed vertebrae of slaughtered cows and used in products such as Jello, seems especially suspect, as it almost certainly contains remnants of the spine and spinal fluid. The average British consumer was incredulous at the idea that blood of possibly infected cows could still be fed to cows destined to be eaten by consumers. (America's Food and Drug Administration still allows cattle to be given feed with the remains of dead horses and pigs, poultry to be fed remains of dead cattle, cattle to be fed remains of dead poultry, and cattle blood to be put into what is fed to American cattle.[23]) A hearing before a congressional committee in March 2001 revealed that mechanical rendering systems for cattle could still transmit parts of spinal chords to hamburger.[24]

Once again, agribusiness in bed with government implored the public to trust the system to safeguard their food, and once again this trust was betrayed.

By the end of 2000, millions of cattle had been destroyed, costing the British beef industry over £4 billion, with the destruction itself costing £27 million. While Britain tried to issue a Final Report assigning blame, something happened across the English Channel that ignited similar fears.

BSE FEARS SPREAD ACROSS EUROPE

In 1996, French prime minister Jacques Chirac had urged the French cattle industry not only to ban slaughtered brains from animal feed but also to ban bovine blood and bone products. But citing a lack of scientific evidence that prions were contained in such products, France followed Britain and allowed use of such animal products in cattle feed to continue. Meanwhile, France continued to defy the EU and banned import of British beef.

The suspect practice of using these animal products in feed is euphemistically called feeding "bone meal" to animals to camouflage the presence of bovine blood, as well as other ground-up parts of dead animals. In blunter terms, "bone meal" is semantic camouflage for ground-up animal remains. How can any food system—any combination of food scientists, public health officers, and federal safety inspectors/bureaucrats—allow such remains to be fed to livestock meant for human consumption and think it safe?

In 1999, French authorities predicted that BSE would be virtually eliminated from its cattle because of the new, safer practices, but a new testing program found thirty BSE-infected cattle in France. Similar tests showed about a hundred cases in 2000.

Then the real bomb hit French papers: three young Frenchmen were infected with vCJD. Moreover, the disease received a human face when seventeen-year-old Arnaud Eboli's emaciated body was shown in a wheelchair on television.[25] In 1997, he had been an athletic 165-pound teenager, when suddenly he started to have unexplained fits (initially diagnosed by psychiatrists as adolescent rage). When a biopsy of his tonsils showed vCJD prions, the worst was confirmed. By the end of 2000, Arnaud was on a feeding tube, paralyzed, and on the verge of death.

Suddenly the reality hit a nation that had always prided itself on the international superiority of its cuisine: it had not been enough to protect French citizens to stand firm on the ban of British beef, and it had not been enough to trust French agricultural and public health officials to ban all possible BSE-transmitting practices in industries within their purview. Now Frenchmen were going to die, and who knew how many more deaths could follow?

Suddenly, boucheries in France suffered 50 to 70 percent declines in business. No one wanted to eat meat anymore. Russia, Poland,

and Italy did the same silly thing France had done and banned French meat, thinking they could protect their citizens with such a ban while not changing their own practices.

When it was announced that fifteen cows were infected with BSE in Germany, despite similar precautions, fears spread there. Portugal and Switzerland had already reported hundreds of cases of infected cattle, and now reports were coming in (no doubt in spite of efforts to suppress them) from Belgium, the Netherlands, Spain, Greece, and Italy. When just a year before it had opposed a ban on bovine blood and gelatin in bone meal, and in response to its angry citizens, Germany now did an about-face.

And Europeans *were* angry. Newspaper editorials and ordinary citizens criticized governments for always being on the side of the cattle industry and never on the side of consumers.[26] Nor did it take these people long to figure out that, if similar dangers should arise from genetically modified food crops, their governments would not act until many people had been made ill and the first human canaries had died.

THE FOUR PERSPECTIVES AND TRANSMISSIBLE ENCEPHALOPATHIES

It is instructive to ask how each of the four perspectives discussed in the previous chapter see the emergence of TSEs in European cattle and humans.

The naturalist response is the easiest to predict: humans strayed too far from traditional, safe, natural ways of farming and started to feed their cows in unnatural ways. Hence, disasters such as mad cow disease were the inevitable result. For theistic naturalists, mad cow disease is punishment for such sins. As G. Cannon writes, "Great epidemics are warning signs, symptoms of disease in society itself."[27]

Some scientists believe that AIDS and HIV infection are many centuries old. Why, then, did they not infect humans until modern times? The answer is that the practices of humans changed, with many more people living in cities (urbanization always increases the transmission of diseases as humans come in closer contact), with people flying quickly over the globe, and with unsafe sexual practices. Society became more impure, corrupt, and perverted, even in regard to something so basic as raising food.

In the same way, naturalists argue, unnatural practices cause transmissible encephalopathies to cross into humans. Exactly how this happened is not the point: the point is that any kind of *cannibalism*, in which animals are eating dead parts of members of their own species, is abhorrent, unsafe, and unclean. In any case, Michael Jacobs says, feeding dead sheep to cattle, or dead cattle to sheep, is "unnatural" and "perverted." He continues: "The present methods of the agricultural industry are fundamentally unsustainable. . . . Risk is not actually about probabilities at all. It's all about the trustworthiness of the institutions which are telling us what the risk is."[28]

Naturalists believe there are natural limits against which humans must not push. Whether ancient Greeks, Jean-Jacques Rousseau, or opponents of suburban sprawl, naturalists think that "less is more"—fewer people, simpler methods, more direct methods, and less complexity.

This attitude is put most bluntly by Vandana Shiva, who evokes the metaphor of a "border crossing" to describe violations of inherently safe, natural practices and ways of creating and raising animals: "The mad cow is a 'border crossing' in industrial agriculture. It is a product of the border crossing between herbivores and carnivores. It is a product of a border crossing between ethical treatment of other beings and violent exploitation of animals to maximize profits and human greed."[29]

For naturalists, the cure for diseases such as TSEs is a return to organic farming, which is more humane than industrial agriculture. Only in this way can trust be restored to our food system.

For naturalists, that humans became infected with mad cow disease was really only to be expected: mad cow disease "is the full story of the apocalyptic phenomena: a deadly disease that has already infected the national cattle herd . . . [and] could in time prove to be the most insidious and lethal contagion since the Black Death."[30]

For progressives, mad cow disease and TSEs are aberrations. They point out that it was politicians, not scientists, who misled the public and lied. Indeed, if not for scientists, no one would know about prions.

Globalists draw a different lesson from mad cow disease in Europe. They see the panic over mad cow disease as an example of the futility of erecting protectionist barriers. What was the first response of France to evidence of human deaths from vCJD? Banning English meat rather than admitting that similar problems could occur in France. Germany did the same thing. Then both countries experienced the same problem in 2000. Then Russia and Italy banned import of French and German meat, undoubtedly while not changing their own practices.

From a globalist perspective, the EU finally came to its senses in late 2000 when it banned bovine blood and spinal products from all animal feed in all EU countries.[31] European officials had finally understood that national trade barriers were not going to get TSEs out of Europe's meat.

Progressives believe that the modern food industry is basically safe and getting safer all the time. Intensive cattle and chicken raising, chemically intensive agriculture, genetic modification of plants—all these are part of a rising tide of plentiful, nutritious, safe food for the world. Where food is plentiful, people can turn to other activities that raise their standard of living and grow the economy.

Progressives emphasize that most fears about TSEs stem from irrational panic and media exaggeration. Given the inundation of print and visual media with bad stories about TSEs in Europe, it is amazing that anyone there at all still eats meat. Progressives stress that the chance of any person becoming sick is less than one in a million. Moreover, they emphasize, one must distinguish between scientists and the bureaucrats, businesspeople, and politicians who *apply* science. Science did not create mad cow disease; bad feeding practices did.

Egalitarians, of course, see mad cow disease as symptomatic of what happens when big business is in bed with big government. Car manufacturers do not rush to inform car owners of defects in their cars and are often only brought to the bar of justice when deaths result in lawsuits. So cattle producers waited until people died before changing their ways. Even then, in 1996, they still wanted to be allowed to use bovine blood in cattle feed—but what do you expect when a desire for profits drives everything? (The reluctance of America to reform the safety of its meat industry, or to stop cannibalistic feeding practices in livestock, similarly points to why the consumer should not trust the meat industry.)

END COMMENT

So does the European experience with TSEs prove that we should all eat organic food? That the food system can't be trusted? That we should follow Europeans in rejecting GM plants?

What we have learned in this chapter is why Europeans suspect their food system and innovations in it. The proper inference is that, if we are in Europe, we should be suspicious of eating sausage, meat, and animal flesh. Or perhaps we should become vegetarians and eat no meat at all.

Nevertheless, the safety issues of recycled dead animals in feed are far different from those of adding a few genes to plants. There is a difference in orders of magnitude of risk here. Indeed, now that we know some of what happens in feeding livestock and creating meat for the table, giving traditional plants a few new genes almost looks innocuous. Given that organic foods are also not perfectly safe, commercially grown plants with genes added to increase vitamin A seem as safe as organic food.

As we shall see in the next chapter, because meat in our industrial food system is filled with dangers, mad cow disease was a predictable kind of problem. But it would be premature to generalize from dangers in meat to dangers in vegetables.

What is not improper to generalize about is trust. Agribusiness is in league with government and politicians from agricultural states to make money. Safety is only regarded because not doing so endangers profits. Safety is not an inherent good pursued by the industry. The experience of mad cow disease, and the continuing problems of safety in meat processing in North America, do show that, should grave problems arise in genetic veggies, agribusiness would be reluctant to publicly expose them. But that does not say that such problems exist or are likely to exist.

5

Is Genetically Modified Food Safe?

It's a very odd thing
As odd as can be
That whatever Miss T. eats
Turns into Miss T.

—*Walter de la Mare*

CASE STUDY: STARLINK CORN

Newspapers reported in September 2000 that StarLink corn, a genetically modified corn approved for feeding to animals but not for humans, had slipped into Taco Bell taco shells that were being sold in American grocery stores.[1] A brand name of Bt corn, StarLink contains a gene for the previously described pesticide, *Bacillus thuringiensis* (Bt), used on crops of organic farmers.

The genes for Bt in StarLink corn create a protein, "Cry9c," that some humans find slightly more difficult to digest than the proteins from traditional corn. After national media revealed the story, about forty people notified the Centers for Disease Control (CDC) and lawyers, complaining of "food poisoning" and other adverse reactions. Samples of their Taco Bell corn shells were taken and frozen, and a six-month study by CDC has so far been inconclusive. Possible lawsuits are at stake.

Besides taco shells, several other products made from corn had inadvertently used StarLink: tortillas, snack chips, and other products made from flour from yellow corn made by Mission Foods Company of Irving, Texas.[2]

Serious concerns arose about this event on two levels: at the level of content, some consumers might suffer indigestion or stomach cramps; at the level of process, organic farmers said this event exemplified why, once grown, the Department of Agriculture (USDA) and Food and Drug Administration (FDA) couldn't prevent humans from eating genetically modified (GM) foods. Both naturalists and egalitarians excoriated the globalistic nature of this food system, because the shells had been made in Mexico and imported by a Texas miller who had also bought traditional corn from farmers in six states and stored them all together, making it difficult to determine one corn from another.

StarLink Corn was developed by Aventis CropScience of the Research Triangle Park, North Carolina. A month later, the Kellogg Company shut down one of its plants in Memphis, Tennessee, after a supplier could not guarantee that its corn was free of StarLink corn. About 9 million bushels of StarLink in the 2000 crop were never accounted for, and presumably had been delivered to more than 350 grain elevators in North America. About 47 million bushels remained on farms and farmers were paid a premium over market price to store them so they would not enter human food markets.

Because food processors had agreed to police themselves to keep new GM foods such as StarLink out of human consumption until further tests could be done by the FDA, this event created ill will among American food producers and Aventis, which was owned by Aventis S. A. of France. Aventis revealed that not all of the two thousand farmers growing StarLink corn had signed agreements to follow procedures keeping the corn out of human consumption.

THE BGH CONTROVERSY: DESIGNER FOOD, ROUND I

In 1975, fears about "Andromeda Strains" (uncontrollable microbes created by recombining DNA) breaking out of supposedly super-safe "P-3 containment" labs created the Asilomar Conference, where scientists imposed a moratorium on gene-splicing experiments. The following year, the National Institutes of Health (NIH) created the Recombinant DNA Advisory Committee, dubbed "the RAC" by researchers impatient with its constraints.

In 1977, Genentech, a leader of the field and a private company, recombined DNA to produce somatostatin, a hormone found in the human brain. The next year, Genentech produced human insulin and licensed it to Eli Lilly. Because no infectious agents could be transmitted from cattle to humans, the laboratory-created, purified insulin was safer than standard, bovine-derived insulin—a fact that then went unappreciated.

As fears diminished, the RAC in 1979 allowed Genentech to use recombinant techniques to produce human growth hormone (HGH), used to increase the height of children suffering from pituitary disorders.

The RAC in the early 1980s also allowed Genentech to create human leukocyte interferon, fibroblast interferon, and pro-insulin. A buying frenzy met Genentech's first offer of public stock. The same meteoric rise in value of stock occurred with Cetus, a major firm that produces enzymes that can form ethylene oxides, which delay the ripening of fruit and thus extend their shelf life.

Genentech soon contracted with the federal government for a vaccine against hoof-and-mouth disease. On another front, it had been known for forty years that cows given BGH produced more milk, but before recombinant techniques, BGH could only be obtained from dead cattle. In 1981, Genentech created a purified, cloned BGH that could be given to cows with a lesser danger of transmission of disease. A similar hormone made chickens grow larger.

The subsequent battle in the 1980s over the use of BGH in dairy cows foreshadowed the battles to come over mad cow disease and genetically modified food.

As "genetic" has shown in "genetically modified food," how a new biological product is named carries great importance for its emotional wallop on ordinary people (cf. "cloning," "test-tube baby," "designer baby"). Those in the dairy industry wanted to use the term "bovine somatostatin" (BST), but activist Jeremy Rifkin called it bovine "growth hormone." The latter sounded sinister, like hormones used by weight lifters. Because of Rifkin's savvy and because scary phrases help the media increase ratings and sell newspapers and magazines, Rifkin's scary phrase won out.

Use of rBGH in the dairy industry opened the door for economies of scale from small mom-and-pop dairy farms to huge dairy farms run by food conglomerates. Monsanto soon bought rights from Genentech to market rBGH to dairy farmers and began to bring it to market.

Naturally, small dairy farms resisted the loss of their income and way of life. They saw their battle in cosmic terms: good (small, family-owned) versus evil (big, owned by stockholders). Globalists applauded the moves toward greater efficiency. Naturalists and egalitarians defended the small dairy farm. Popular singers staged fund-raising concerts, which amounted to little more than sticking a finger in a broken dike.

The FDA required Monsanto to prove that milk produced from cows given rBGH was just as safe as traditional milk; the resulting research showed that the new milk had small traces of growth hormone, similar to that in traditional milk. Moreover, humans digested all residues of rBGH. In one crucial concession to critics, the FDA allowed states to mandate labeling of milk derived from cows given rBGH, leaving the door open for boycotts.

It took a long time, a decade, for Monsanto to jump through the regulatory hoops to bring Posilac to market. Progressives emphasize that, as with genetically modified foods, this shows how much capital, time, and research is needed in America before any such food is allowed on the market. Indeed, there's an interesting dilemma here: activists such as Rifkin carp that large conglomerates control more and more of our drugs and biotechnology, but because of critics such as Rifkin, only large corporations have the resources to battle critics for decades to bring new products to market and, after all that, to pay damages from suits if anything goes wrong. When small companies try to do this, they can't meet payroll and may go bankrupt (see Calgene below).

So it took over a decade to bring rBGH to market: it was created in 1981, Monsanto began to get it certified for the market in 1984, FDA granted approval in 1993, and it was first used commercially in America in 1994. If you were a small company with ten investors who in 1979 had each put up $50,000, none of them would've made a dime for fifteen years. Sometimes critics forget that 75 percent of Americans own stock in such companies, especially in their retirement plans.

Like GM foods, rBGH faced opposition that delayed introduction of the new milk into grocery stores. Monsanto and other agribusinesses such as Cargill were shocked by the emergence of egalitarian and naturalist activists who urged boycotts of rBGH milk.

Jeremy Rifkin's Foundation for Economic Trends started its Pure Food Campaign. Other activists, such as the Wisconsin Family Farm Defenders, focused on saving ways of life. Since everyone ate food, elitist environmental organizations such as Greenpeace sought to get more public support by scaring people about GM food. Consumers Union, which historically suspects big business, joined the fray. Animal rights groups such as People for the Ethical Treatment of Animals deplored the conditions of dairy cows given rBGH. Previewing the coming fight over GM food, many naturalist and egalitarian groups tried to rally North Americans around opposition to rBGH.

Behind the attack on the safety of such milk lay a different kind of issue, one seen also in the opposition of European and Japanese farmers to importing American crops. "Some analysts predicted that the [rBGH] technology would accelerate a shift from small to large dairies and from the Northeast to the South and Southwest as principal production regions [of milk]."[3] So concerns about rBGH may mask economic battles about whether Wisconsin or Vermont will dominate milk production.

As naturalist Vandana Shiva says about this controversy, using genetic-created BGH turns cows into milk-producing machines and makes them part of industrialized dairies. She wants instead India's traditional system, which emphasizes "women's traditional role in caring for cows and processing milk" and in which recycled cow manure is integral to the country's millions of small farms.[4]

For progressives, all this opposition occurred irrationally because no evidence showed that milk produced from rBGH given to cows was unsafe. For progressives and globalists, the opposing groups did not respond to a scientific estimate of the danger of rBGH milk.

Of relevance to the controversy over labeling GM food, and because of the prior FDA decision, a coalition of small northeastern dairy producers successfully created fears among consumers about use of rBGH in dairy cows, and hence won passage of laws in several states requiring labeling of rBGH milk in grocery stores. In these states (Wisconsin and the three most-northern New England states), production of rBGH milk dropped to the lowest levels in the country.

JEREMY RIFKIN

We cannot discuss the safety of GM food without discussing activist Jeremy Rifkin and his media-savvy tactics. Rifkin scares people about biotechnology and genetically enhanced food. Rifkin and his ilk create scary names such as "FRANKENFOODS" and "MUTANT FOODS."[5] In a book with Ted Howard called *Who Should Play God?*, Rifkin mastered emotionally alarming phrases that got the media's attention, although one reviewer correctly dubbed the book "pseudoscientific blather."[6]

For over thirty years, this self-styled gadfly has criticized scientific, medical, and biotechnological advances. The bête noire of scientists, who despise him, he opposed in vitro fertilization in the 1970s for infertile couples. Creating "test-tube babies," he intimated, wrongly created babies in artificial ways, as if out of chemicals, not human sperm and egg.

Both naturalist and antiglobalist egalitarian, Rifkin predicted the end of meaningful work in North America (unfortunately for him, just when the economy soared in the 1980s and created millions of good jobs in technology and medicine).[7] Two decades after bioethicist Peter Singer published *Animal Liberation*, Rifkin unsuccessfully tried to sell the same issue.[8] In 1979, he celebrated fundamentalist religion as a liberating force that could overthrow materialistic ethics and set the stage for a new, spiritual reawakening in the twenty-first century.[9] This prediction, like so many of his others, did not come to pass.

In 1983, he got sixty religious leaders, from Jerry Falwell to twenty-two Roman Catholic bishops, to sign a resolution asking Congress to ban creation by recombinant DNA of human reproductive cells. He later opposed trials in humans for gene therapy against cancer. When asked why he fought against the first use of gene therapy, he not only claimed the scientists weren't ready, but that "Suffering is part of every species' existence."[10] Fatalistic acceptance of disease, he implied, should be part of human wisdom, and it is hubris for physicians to try to cure it. In this attitude, he mirrors Leon Kass, the bioethicist appointed by President Bush to chair a committee approving research from embryonic stem cells, who recently has told us that we should not cure death even if we could. (Why? Death is good for us, helping us love our children and develop true virtues.)[11]

Progressives and globalists track Rifkin's moves. Matthew Hoffman of the Competitive Enterprise Institute in Washington, D.C., characterizes Rifkin as "a very unscrupulous man, a sophist, . . . a master media manipulator, [who] loves to make all sorts of hysterical contentions or scary predictions which the press picks up on. . . . He uses the regulatory process and the courts to slow the advance of technology. If there is anything he hates, it is technology."[12]

In 1997, Rifkin published *The Biotech Century*, in which he predicted the globalization of pharmaceutical, seed, agricultural, and medical companies, led by Astra-Zeneca, Monsanto, Novartis, and Cargill.[13] His egalitarian, almost Marxist manifesto warns that these companies exploit poor people and that nothing good comes of their efforts. He warns of the dangers of genetically modified food. He opposes patenting human genes, implying that the essence of humans was being commercialized.

In the late 1980s, he launched a major campaign against use of cadaver-derived BGH in dairy cows (it was not until 1994 that rBGH was used). A progressive government official, Mike Phillips of the now-defunct Office of Technology Assessment, says Rifkin likes to force his "opinion on how things should be in this world, no matter what the facts are. He makes use of the fears of others."[14]

In sum, Rifkin has been called antimedicine, antiscience, antibusiness, and a socialist. He has been accused of wanting to save the family farm at all costs and wanting a return to primitivism. However, it is difficult to know whether he really believes any of this. All we know for sure is he aggressively promotes his own sensationalist books and that every other year he will have a new alarmist book on the market.

THE ICE-MINUS CASE

With the next story, we move closer to why Rifkin matters to evaluating the safety of GM food. In 1983, Rifkin and others sued the University of California at Berkeley to stop field tests of the genetically modified bacteria, *Pseudomonas syringae*, that would lay this bacteria on top of potatoes to stall (by a few degrees temperature) formation of damaging frost. Federal judge John Sirica (who presided over the trial of the Watergate conspirators) heard this precedent-setting "ice-minus case" ("potatoes minus ice").

In this case, naturalists and egalitarians vanquished agribusiness and biotechnology, putting the latter on notice. Rifkin won by spotting a weakness in federal regulations and by pushing hard on it. Several well-known ecologists, including Peter Raven, head of St. Louis's famous Missouri Botanical Garden, testified that effects of release of such bacteria on California's ecosystem could not be predicted. The coup de grâce was when Rifkin et al. argued that the government had not shown any track record of evaluating safety in genetically modified plants and had no regulations or agency in place to spot emerging dangers. Congressman Albert Gore soon held congressional hearings on the environmental implications of genetically enhanced crops.

This suit and the subsequent hearing, as well as the subsequent classification by the FDA of this bacteria as a pesticide, created unexpected, formidable expenses. Henceforth, such crops would need to be carefully (excessively?) monitored and tested, as if they were something categorically new in nature. The final effect of all these new hurdles was that the bacteria's owners gave up and never conducted field trials.

This suit had real costs to the public. Subsequent freezes in 1990 and 1998 in California cost citrus growers between $600 and $800 million.[15] The ice-minus bacteria almost certainly would have prevented some of this damage. As a result, prices paid by consumers rose for citrus fruit.

FLAVR SAVR TOMATOES

In the early 1990s, Rifkin launched another campaign, against another genetically modified food. His "Pure Food Campaign" targeted the "Flavr Savr" tomato made by Calgene of Davis, California. "We'll win on Pure Food within the next year," he said in 1992. "We'll defeat the Flavr Savr just as we defeated BGH," he gloated. "We will [also] prevail on bio-engineered food, though it may take a little longer."[16]

Flavr Savr tomatoes contained so-called antisense RNA, which inhibited expression of a gene that causes tomatoes to soften. The resulting fruit had value for both growers and consumers by producing tomatoes that were plump, sweet, juicy, and slow to rot. Because the fruit would stay firm longer, it could be harvested mechanically without bruising and could be transported over long distances. Military personnel in northern Alaska in the middle of winter could have tomatoes flown in that tasted like those off the vine in spring.

This could also have been the breakthrough product that allowed consumers to accept GM fruits and vegetables. Consumers generally hated tomatoes that had been picked green and ripened in the warehouse with ethylene gas. This process made the tomatoes look red, when in fact they lacked all taste. But if the tomatoes had started to turn red before shipment, they would arrive rotten or rot within a day at the store.

Taking a cue from the way Scientologists use movie stars as shills to legitimize Scientology, Rifkin enlisted several gourmet chefs to vow that they would not use genetically modified foods.

In this campaign, Rifkin failed legally but still won. The FDA in 1994 declared the Flavr Savr safe, allowed it to be grown, and did not require they be labeled as different. (The same year, the FDA allowed use of genetically enhanced and genetically purified chymosin, a key enzyme used in producing cheese.) Party because of the delays and uncertainties caused by Rifkin's campaign, the Flavr Savr tomato proved too costly to produce and was never commercially successful.

There is another side to the failure of Flavr Savr.[17] Businessmen within Calgene hyped the tomato at the earliest date on the scantiest of evidence. True, they had evidence that the tomato would survive on the shelf and taste good weeks later, but they did not have evidence that it would survive shipping across the country and taste as good. Like the run-up of technology stocks on the NASDAQ, investors bought up the stock based on the hope that quick science would solve the remaining issues. That science didn't happen quickly enough to make money, and Monsanto took over Calgene in 1997 after it almost went bankrupt.[18]

SUPER-BROCCOLI

Consider carefully the following article carried by many news agencies in the spring of 2000:

Super-Broccoli Bred to Prevent Cancer

May 25 2000. Super-Broccoli bred from garden broccoli and a wild Sicilian variety is the latest veggie to hold out anti-cancer promise. The team that developed super-broccoli at the John Innes Centre at

> the Institute of Food Research in Norwich, England now have two commercial partners.
>
> "The super-broccoli looks and tastes the same as ordinary broccoli," says Gary Williamson, a member of the research team that bred the plant. Compared with regular broccoli, super-broccoli contains 10 to 100 times as much sulphoraphane, the substance that helps to neutralize cancer-causing agents in the gut. . . .
>
> The researchers are at pains to stress that super-broccoli is not a genetically modified (GM) plant. "No gene has been inserted through genetic modification," Richard Mithen, a research team member, told Reuters. "This is classical breeding. But we speeded that breeding program up by using DNA fingerprinting technology."[19]

This article was posted on many web sites by organizations devoted to health foods and natural eating. Note that the creators of super-broccoli take pains to claim it was not "genetically modified," although they admit that they used "DNA fingerprinting technology," a less objectionable-sounding term.

What is really remarkable is that if we modified broccoli by carefully adding a few genes to produce a hundred times more sulphoraphane, it would be called "genetically modified" and then be subjected to all the field trials, hysteria, and regulatory burden that plagued ice-minus potatoes and the Flavr Savr tomato.

However, traditional crossbreeding techniques would create no such hysteria, testing, or regulation. But crossbreeding introduces hundreds of genes into traditional broccoli from the wild Sicilian variety. How do we know that something terrible is not lurking in these new, wild genes? The best answer is that, "Well, broccoli is still broccoli, crossed with wild broccoli genes or not."

Now consider a traditional broccoli where there are not, say, one thousand unknown genes and resulting unknown proteins, but just a few new genes that produce known, tested-for proteins. And this new produce is extensively tested on rats, mammals, and human volunteers before it's introduced to the general public. Which is more dangerous, by far?

NEW ALLERGENS?

"Genetically engineered foods have the potential to introduce new allergens and toxins into the food supply," claimed activist Joseph Mendelson at a press conference during a protest outside the Food and Drug Administration in Washington, D.C.[20] He directs legal matters for the Center for Food Safety, a naturalist PR firm funded by organic farming (not to be confused with the FDA's Center for Food Safety and Applied Nutrition).

Inability to easily eat certain foods does not mean one has an *allergy*. More likely, one is merely *food intolerant*. Allergic reactions involve the immune system and release of histamines, whereas the more benign food intolerance means digestion is difficult, often for lack of a particular enzyme (my father always had such problems from eating ice cream, but he never stopped eating this food because he loved it. People can easily continue to eat a food to which they are "intolerant").

The difference between allergies and food intolerances matters to the debate about GM food. Only 1 percent of adults suffer from true food allergies,[21] although two out of five Americans believe they are allergic to some food.[22] Over 90 percent of allergic reactions come from known proteins in peanuts, wheat, soybeans, or cow's milk. Over a hundred thousand North Americans can't eat grains because their digestive tracts cannot absorb gluten, which is found in products made from wheat, barley, and oats (rice, however, is free of gluten). The other 10 percent of allergies come from proteins in berries, shellfish, corn, beans, and gum arabic.

A much higher percent of people cannot tolerate some foods; indeed, most of us are probably intolerant of food of some kind. Symptoms of food intolerance sometimes resemble allergic reactions: skin rashes and hives, difficulty breathing, indigestion, vomiting, diarrhea, intestinal pain. Indeed, 70 percent of the planet becomes intolerant to lactose after childhood, even though they can process it easily as infants and children.[23]

INTOLERANCE TO STARLINK CORN?

Almost all known food allergies are due to proteins. As discussed above, scientists know which proteins commonly cause true allergic reactions.

For naturalists, the index case of allergens with GM foods occurred when the University of Nebraska at Lincoln—to create more protein—inserted a gene from Brazil nuts into soybeans. Predictably, people with nut allergies could not eat such soybeans, and many products are made from soybeans. Testing the new soybeans showed they were allergenic, so they were not brought to market.

It is difficult to understand why naturalists and critics of GM food so often cite this case, because of the predictability of known allergens and because the system *worked:* no humans ever had bad reactions because the new soybeans were never introduced. This is like criticizing public health officials who prevented the reemergence of smallpox by arguing that, nevertheless, the disease might have harmed people if the officials hadn't acted.

As stated at the beginning of this chapter, the Cry9C protein in StarLink (Bt) corn may take longer than normal to break down in the human gut, and sometimes, when it takes a long time for gastric juices to break down a protein, the lengthy process creates an allergic response in humans. Because it was not known that Cry9C did not create this sort of response, it was not ruled out as an allergen to humans.

In other words, the protein was known to be neither dangerous nor safe. But foods with unknown allergenicity are not normally released to the nation.

The Brazil-nut-gene-in-the soybean case certainly proved that allergic effects can be transferred to new crops via transfer of genes. If any such produce were to come to market, it clearly should be labeled so that consumers with such allergies could avoid it.

But labeling foods never known to create allergic reactions, but which have some new genes, is a different problem. Naturalists claim that people with rare food intolerances or allergies will unknowingly be exposed to risk if new genes are inserted into old foods. Indeed, it is almost certain that some people will experience unexpected food intolerances, perhaps even allergic reactions, to some foods with new genes. Moreover, naturalists argue, these will only be the *immediate* effects of the new genes. If allergic responses are the early-warning system of the immune system, what about long-term effects? Naturalists have a strong point here, arguing that the public is being used as guinea pigs in a longitudinal experiment about the safety of new food.

Progressives report that you cannot blame a system, when it *prevented* such food from getting to the market, as the real source of the problem. The new soybeans were not brought to market and nobody was harmed because of scientific testing and because the preventive system worked.

Nevertheless, progressives must be embarrassed by the failure of the system to prevent StarLink corn and its Cry9c protein from entering human food. Such corn may have caused food intolerance in some people and should not have been released until testing proved it safe. StarLink corn was only supposed to be fed to animals, not humans. Moreover, rules were too loose, as not all farmers knew about keeping StarLink corn segregated from traditional corn.[24]

Egalitarians claim that it was just too profitable for farmers and distributors not to slip the new corn in with the old: where was the profit in segregating the corn? The subsequent difficulties of tracking the StarLink corn, in a hopeless effort to keep it from entering human food, showed the disadvantages of a national/international distribution system premised on blending all sources together to create one uniform product.

DID ARPAD PUSZTAI'S RATS SHOW GM POTATOES UNSAFE?

In 1998, Arpad Pusztai, a sixty-eight-year-old senior professor at the Rowett Institute in Aberdeen, Scotland, made big news by claiming that rats developed tissue damage after being fed genetically enhanced potatoes, but did not develop such damage when fed traditional potatoes. In an extraordinary development in medicine, these claims were first published not in a plant journal, but in a British medical journal that had previously been held in high regard, the *Lancet.*

What did the *Lancet*'s editors know about such studies? Not much, it turned out. The editors claimed the issues needed to be publicly debated, so they were publishing the piece without the usual peer review by plant and animal scientists. One wonders if they will similarly allow medical articles to be publicly debated on astrology, psychic surgery, and therapeutic touch.

Pusztai and Stanley Ewen, a pathologist at the University of Aberdeen, fed six rats potatoes altered to contain a gene that makes the protein lectin, which increases the potato's resistance to damage from insects and worms. For controls, they fed six other rats traditional potatoes and fed another six rats potatoes laced with lectin.

In only ten days, the two scientists claimed, the intestines of the rats fed the genetically inserted lectin showed a very uneven pattern of thin and thick spots, suggesting damage. Friends of the Earth leaped to praise the study and applauded the *Lancet* for taking the unusual step of publishing it.

But the Royal Society, Britain's parallel to the American National Academy of Sciences, criticized Pusztai's study, claiming it contained too few rats and that the results did not, and could not, prove what he claimed. The Royal Society also condemned the *Lancet* for publishing a study that had not been reviewed by Pusztai's peers.

In early 1998, Pusztai claimed on television that he had fed five rats for 110 days on potatoes into which genes from the snowdrop protein and jackbean had been inserted and that these rats subsequently had damage to their immune systems and internal organs. Although this seemed like very strong evidence, Pusztai lacked the data to back up his claims. After a five-day investigation by his colleagues, Pusztai was suspended from the institute for making claims unproven by any evidence. A senior scientist said the Rowett Institute was deeply embarrassed by the incident.[25]

Chinese researcher Zhang-Liang Chen of Beijing University attempted to duplicated Pusztai's results in 2001, but he found that in over forty studies no damage had occurred to hundreds of rats fed genetically modified foods.[26] Chen's rats, coerced into being vegetarians, feasted on raw sweet peppers and tomatoes enhanced with a gene from the cucumber mosaic virus. Comparing their internal organs, blood, and sperm with controls revealed no differences.

False reports such as those by Pusztai harm many people. Visual and print media sensationalize possibly bad news and never similarly sensationalize corrections. (Imagine this headline in your favorite stately newspaper: "WE SCREWED UP!") Urban legends are created when a rumor is passed on as a truth, and the messages snowball, such that years (or even centuries!) later, the rumor still circulates.[27]

Damage is worse when the initial rumor is not a complete falsehood but a news article on a scientist reporting about something as basic as food. So it is no surprise that eighteen months after Pusztai's claims, and after the claims had been widely retracted by his Rowett Institute, they influenced people everywhere. As one gardening editor wrote in her nationally syndicated Scripps-Howard newspaper column, "Studies are beginning to suggest that genetically altered potatoes may cause brain and liver damage in rats."[28] Newsgroups on the Internet widely report that "rats were damaged by eating genetically modified potatoes."[29]

SOME GM FOODS ARE SAFER THAN TRADITIONAL FOODS

As discussed, genetically engineered bovine growth hormone (rBGH) contains little possibility of infection with contaminated prions, viruses, or bacteria, unlike "naturally derived" BGH. Similarly, genetically modified Chymosin, an enzyme that helps milk coagulate in the production of cheese, used to be obtained from the stomachs of calves. Like Factor 8 clotting factor or rBGH, cloned chymosin is safer than the traditional form and allows moral vegetarians to feel good about eating cheese.[30]

A second argument in favor of safety is that genetically altered crops are extensively tested, whereas new, crossbred variations of traditional crops are not. For example, Monsanto subjected Roundup Ready soybeans to eighteen hundred analyses comparing the new soybeans to traditional ones for hundreds of substances and effects, including proteins and fatty acids. Monsanto concluded that the two kinds of soybeans did not differ in any way.[31]

To prove safety to people, Monsanto fed high dosages of the enzyme made by the new soybeans to rats, chickens, and cows, with effects on animals identical to those produced by traditional soybeans. After human tests, the FDA also found that enzymes of both new and old soybeans digest in human stomachs within fifteen seconds.

"We know more about the safety of Roundup soybeans than almost anything else we eat," said Anthony Trewavas, a professor at the University of Edinburgh.[32] Plant biochemists have a lot of experience with proteins and their tests show that Roundup Ready soybeans do not create dangerous new chemicals but create the same kind of proteins as traditional soybeans. Trewavas also points out that traditional crops haven't been tested at all, even though most such plants produce natural toxins and even though some new varieties of potato and celery created by Mendelian methods have made people sick.

This is certainly worth emphasizing. Most people probably eat processed foods, such as fast foods and artificial creamers, that have received far less testing that Roundup soybeans. They also try new foods and spices in exotic restaurants (Ethiopian, Indonesian, Iranian), yet most such foods and spices have not been as extensively tested as GM veggies.

Also, we have to trust someone, sometime. I see people ordering organic foods in a local restaurant but ordering Diet Cokes. Well, Diet Cokes contain saccharine, and the same scientists they distrust about GM food certified saccharine as safe after tests in rats. You can't pick and choose when you trust scientific results.

SYNERGY

For naturalists, the modern food system created by international agribusiness has grown so complicated that unintended harms will occur just because of complexity: too many small details must work just right for the final product to be safe. The system is fragile, vulnerable to infection, and does not have enough redundant safeguards. Naturalists point, of course, to the transmissible spongiform encephalopathies (TSEs).

A British manufacturer once used a sugar substitute in hazelnut yogurt, causing an outbreak of *C. botulinum* that killed one person and made another twenty-seven people very ill. The substitution of an artificial sweetener for sugar inadvertently allowed spores of *C. botulinum* to germinate and to produce toxins, whereas with sugar, such growth was inhibited. As Nichols Fox writes,

> Because of an apparently insignificant change in the manufacturing process, a previously safe product became a very dangerous one. The mistake was in seeing the yogurt as simply a collection of ingredients, each acting independently. The synergistic effect of the sugar, a seemingly unimportant ingredient, on the whole product hadn't been considered.[33]

Synergy refers to the combined, overall effect of many separate parts, which may be more than the sum of the parts. Despite appearances to the contrary, such parts may not be interchangeable in their subtle effects on the whole.

Napoleon once offered a prize to the scientist who could create a method of preserving food for his armies; Nicolas Alpert won the prize by proving that this could be done by heating food and then sealing it airtight. He invented preserving food in jars and cans. But at the time, no one knew much about lead poisoning, which could result from solder used to seal tin-coated steel cans. Napoleonic troops who ate out of such cans got sick and even died. (When ice crushed the ship of Sir John Franklin, who was seeking the Northwest Passage in the mid-nineteenth century, his men died not from exposure but from lead poisoning from using such cans.)[34]

The argument from synergy about GM foods is easy to follow. A small change in the composition of a traditional food can create unforeseen, bad effects as a result of the interaction of this small change with a complicated system. Adding new genes to make a rabbit glow in the dark, a tomato soften more slowly, or lettuce resist frost may seem benign in themselves, but who can predict what inadvertent disasters may later ensue if we continue such unnatural combinations?

The infamous example of tryptophans showed how toxic results could result when food is produced in another part of the world, as often occurs in a global food economy. Here is an example from naturalist food author Nichols Fox's *Spoiled: Why Our Food Is Making Us Sick and What We Can Do about It:*

> In 1989 public health officials in Minnesota spotted an outbreak of eosinophilia-myalgia syndrome, which can be fatal, and connected it to a product containing tryptophan, which some patrons of health-food stores had been buying as an apparently safe relaxant. After one death, the FDA re-called the product from the market. The investigation pointed to one Japanese company as the source of the problem. The company used a fermentation process involving *Bacillus amyloliquefaciens* to produce tryptophan, but two factors had changed in the production: By reducing the amount of powdered carbon used in purification, the company had introduced a new genetically altered bacterial strain of *Bacillus amyloliquefaciens* that differed only in its enhanced ability to synthesize two chemicals. This change apparently caused more of the agent responsible for the outbreak to remain, thereby creating a dangerous product that resulted in illness and death half a world away.[35]

Consider another example of bad synergy: how prepackaged chicken may drip on a supermarket's conveyor belt, where the next customer places fresh apples and celery. The unobservant consumer takes home her fruits and vegetables and forgets to wash them, thereby possibly ingesting salmonella. Consider the salad bars in fast-food chains, run by teenagers indifferent to standards of food cleanliness, and who often fail to wash their hands: an ideal place for unsafe plants to infect people.

Rebecca Goldberg of an anti-GM food organization, the Environmental Defense Fund, and a member of the National Academy of Sciences committee investigating GM crops, argues that the process of inserting new genes is not as precise as progressives say. "Genetic engineers still can't control very well where a gene is inserted or how many copies are inserted," she says. "And if you plant a new gene in the middle of some existing genetic material, you can screw up the function or change the way the genes work."[36] According to Goldberg, when an experimental plot of genetically modified petunias was planted in Germany in 1990 in an attempt to get some white petunias, many more turned white than expected and, when temperatures dropped, these turned to variegated shades. "This was supposed to be a very straightforward type of genetic engineering," she says, "but it didn't turn out that way."

It is precisely this kind of unexpected effect at the genetic level that lights fires under critics. *Bon Appetit* magazine worried that "it's possible that genetic manipulation could enhance natural plant toxins in unexpected ways—by switching on a gene that, in addition to having the desired effect, pumped out more poison." Responding to a query about such dangers of GM crops to humans, Paul Billings, a Texas medical geneticist well known in bioethics for his criticisms of big insurance companies, also suspects ag/pharm conglomerates. "We'd never know [of any new dangers] until people started dropping," he told *Bon Appetit.*

EATING MEAT: A COMPARATIVE STANDARD OF RISK

To establish the safety of genetically modified food, some generally accepted standard of safety must be used. Too many critics imply that genetically modified food must be risk free. The first chapter exploded the myth that organic food is risk free while genetically modified food is high risk. But how high is the risk of GM food? What standard of comparison should we use?

Well, most people eat meat and even if they feel eating meat is wrong or unhealthy, most do not think it *unsafe* to do so. Whether this belief is true or not is a different question. For now, all that matters is that most people allow their children and grandparents to eat all kinds of meat, eat it themselves, and trust our system of food to deliver it to them safely. If the same system can do the same or better with GM plants, these same people should have no problem buying and eating GM veggies.

Now most Americans eat meat created for and sold within the fast food industry, such as hamburger, as well as processed meat sold for luncheon meats, such as bologna and salami. On an average week, the typical American eats three hamburgers, and in an average month her kids aged three to nine eat at MacDonald's.[37] Probably as many Americans eat sandwiches containing processed ham, salami, and bologna. Americans spent $110 billion in 2000 in fast-food restaurants, more than they spent on higher education, personal computers, or new cars.[38]

How, safe, then, is meat in general? The answer is controversial. Ever since Upton Sinclair in 1906 exposed the horrendous practices of Chicago meatpacking firms in *The Jungle*, which led President Theodore Roosevelt and Congress to require the FDA to supervise the slaughter and preparation of meat in America, the practice of killing and eating dairy cows, steers, hogs, and chickens has been riddled with safety problems. The meat industry claims that it put behind it long ago the brutal, unsanitary practices described in Sinclair's book, but is this true?

The answer cannot be separated from the vast changes that have occurred in American eating. The fact that so many Americans eat at fast-food restaurants—a category that includes high-end franchises such as Outback Steakhouses, Johnny Rockets, Fuddruckers, and Chili's—means these enterprises made the meat industry meet their needs. Above all, fast-cooked meat needs to be standardized and available in vast quantities on demand. Over several decades, these needs have created the largest industrial food system in history, one stretching not just over North America, but over the entire planet.

In practice, this means that a dozen huge meat companies control the feed lots, slaughterhouses, and packing plants for all fast-food restaurants that serve meat in America. In practice, this means creating huge, forty-thousand-pound lots of hamburger that come from hundreds of different dairy cows.[39] In practice, this means feeding lots containing 100,000 cows, all given antibiotics and anabolic steroids to increase weight while they consume three thousand pounds of feed to gain four hundred pounds immediately prior to slaughter.[40] In practice, this means assembly lines for killing at rates as high as three hundred cows per hour, where the lines can be legally stopped for safety only if an inspector smells or sees a rotten carcass. In practice, this means mechanical rendering systems that cut into the spinal chord (which means if BSE ever broke out in America, it could easily be transmitted to big lots of beef). In practice, it means feces on feet and hides easily contaminate the meat. In practice, this means that *E. coli* infects most hamburger lots, such that the industry takes the position that it is the consumer's duty to cook hamburger well to prevent sickness, not the industry's duty to sell infection-free meat.

"Under current regulations," asserted Dr. Russell Cross, chief of the meat industry's Food Safety and Inspection system, "the presence of bacteria in raw meat, including *E. coli* O157:H7, although undesirable, is unavoidable, and not [itself] cause for condemnation of the product. Because warm-blooded animals naturally carry bacteria in their intestines, it is not uncommon to find bacteria on raw meat."[41] In other words, given this mechanized industrial meat-production system, infected products are inevitable.

Which may be acceptable if you're cooking your own meat, but what if the eighteen-year-old in the local fast-food restaurant gets sloppy cooking the hundreds of burgers he flips all day? Who gets sick then?

If hamburger were made the old-fashioned way, by local butchers who bought meat from a local farm, mass infection of hamburger by *E. coli* would not be a persistent problem. It is the vast, assembly-line nature of our system of producing meat that leaves it open to such risks. (At the same time, it must be mentioned that such standardization theoretically also allows uniform public health practices to govern all aspects of the operation and, if followed and really monitored by inspectors, could produce a reasonably safe system.)

When the famous outbreak of the lethal *E. coli* O157:H7 occurred in 1993 in Jack-in-the-Box franchises in the Northwest, resulting in the hospitalization of thirteen children, five in intensive care, a batch of hamburger from the Vons Company of Los Angeles had created one infected lot of *seventy-seven thousand* hamburger patties. Obviously, all the patties had to be recalled, which was difficult, because they had been shipped all over the West.

In 2001, a hidden video camera in a slaughterhouse in Washington State revealed live and thrashing cattle chained upside down, moving down a processing line; affidavits by workers claimed that between 10 and 30 percent of cattle in this plant were processed while still conscious.[42] Federal law requires animals to be unconscious when they are cut up. If such laws about pain are being violated, one can imagine how many laws about hygiene are also violated.

It goes without saying that vast operations that process cattle and hogs create enormous amounts of waste, especially lagoons filled with urine and feces, the stench of which can be smelled miles away. Greeley, Colorado; Dakota City, Nebraska; and the countryside of Iowa are now infamous for the environmental hazards of their industrial beef production facilities. In North Carolina, Iowa, and Nebraska, similarly vast operations contain a million hogs, producing eighteen hundred pounds of noxious hydrogen sulfide a day.

Keeping such products safe and clean is very difficult. More important, in such vast hog and cattle operations, should one animal get sick, sickness quickly spreads to thousands of others. For this reason, and because, strangely, it causes animals to gain weight, most cattle, hogs, and chickens get massive antibiotics in their feed. How much of these antibiotics are passed on to humans is unknown, but their regular and massive use at the very least creates more resistant bacteria.

This new industrial food system took beef- and meat-processing away from the urban Midwest (especially Chicago) and its union labor, dispersing it to rural areas where nonunion, immigrant (sometimes illegal) workers cut and slice animals in one of the most dangerous jobs in America.[43] Similarly, the eight chicken processors that control two-thirds of the market "shifted almost all their production to the rural South, where the weather tends to be mild, the workforce is poor, unions are weak, and farmers are desperate to find some way of staying on their land. Alabama, Arkansas, Georgia, and Mississippi now produce half the chickens raised in the United States."[44]

This vast new industrial food system has caused a rise in incidents of food sicknesses over the past decade, a stealth phenomenon that most people choose to ignore until they get sick:

> The Centers for Disease Control and Prevention (CDC) released today the most complete estimate to date on the incidence of foodborne disease in the United States. According to data published in the current issue of CDC's *Emerging Infectious Diseases*, CDC's peer-reviewed journal that tracks new and reemerging infectious diseases worldwide, diseases caused by food may cause an estimated 325,000 serious illnesses resulting in hospitalizations, 76 million cases of gastrointestinal illnesses, and 5,000 deaths each year.[45]

All these illnesses and deaths occur in our traditional, safe industrial food system. Moreover, many infectious agents other than *E. coli* OH157:H7 make people sick, including *Campylobacter jejuni, Crytosporidium parvum, Cyclospora cayetanesis, Listeria monocytogenes*, and Norwalk-like viruses. Salmonella, listeria, and *E. coli* microbes are in many cattle and chicken.[46]

So if infected meat is a problem in the meat industry, is there an adequate system of inspection to weed out rotten or infected meat? There is not, and this has created a continuing scandal in the American food system.

The meat industry has steadily resisted federal oversight and real inspections.

> During the 1980s, as the risks of widespread contamination increased, the meatpacking industry blocked the use of microbial testing in the federal meat inspection program. A panel appointed by the National Academy of Sciences warned in 1985 that the nation's meat inspection program was hopelessly outdated, still relying on visual and olfactory clues to find disease while dangerous pathogens slipped past undetected. Three years later, another National Academy of Sciences panel warned that the nation's public health infrastructure was in serious disarray, limiting its ability to track or prevent the spread of newly emerging pathogens.[47]

The basic problem is that congressmen, and especially senators, from meat-producing states block any serious efforts at reform. The wife of Senator Phil Gramm of Texas sits on the board of directors of IBP, the largest meatpacking company in the world.[48] Jesse Helms of North Carolina, Mitch McConnell of Kentucky, and Orrin Hatch of Utah enjoy the financial backing of the meatpacking industry.

The reductio ad absurdum of our system is that USDA meat inspectors cannot legally order a recall of a lot of meat they believe to be infected. Whereas a defective toy can be recalled, defective meat cannot. Meat inspectors can only "consult" with the company and "suggest" that the meat be recalled. Similarly, once a company decides to pull an infected lot from the public, it is under no legal obligation to inform the public of the infected batch. Nor does or must the USDA inform the public of a recall of infected meat from a fast-food restaurant.[49]

To any rational onlooker, the American system of meat inspection is ludicrous. It is fragmented and has no overarching federal agency insuring public health. The FDA regulates eggs, but the Department of Agriculture regulates chickens, while the CDC tracks diseases of threat to humans, but has no legal authority to intervene. The secretary of agriculture is a political appointee and usually from the meat or agriculture industry. The fox guards the hen house.

Congress adopted our system of inspecting meat after several scandals over rotten meat. Our ninety-year-old system only allows meat inspectors to stop the assembly line if they notice a completely rotten carcass. Anything less putrid must be allowed to pass.

And, it seems, meat inspectors can't even stop the line when they actually spot rotten meat. After a few courageous inspectors in 2000 blew the whistle on rotten meat, the USDA tried to shut down the plants involved, but a Texas judge ruled that they could not, even though the plant had illegally high rates of salmonella bacteria. In addition, companies always seem to know in advance when inspections are coming, so factories can be cleaned in advance, and this practice has continued for decades.

As a result of this mess, meat inspectors have no real job anymore. Positions have not been funded and many inspectors have retired early or quit. In 1978, there were twelve thousand inspectors, but in 2000 there were only seventy-five hundred. Bills giving the USDA the authority to order recalls, and to boost the authority of meat inspectors, were defeated in Congress in 1996, 1997, 1998, and 1999. With a new president in 2000 from Texas, a beef state, chances for reform look poor.

OTHER COMPARISONS

The meat industry hopes its salvation lies in irradiation, which does not kill bacteria but disrupts genes of bacteria so they cannot reproduce. In a move parallel to the case of genetically modified plants, the meat industry opposes mandatory labeling of irradiated meat and even wants to lie by calling the process "cold pasteurization."[50] One critic opposes irradiation because he does not believe largely illiterate workers can safely handle complex electromagnetic and nuclear technology. Moreover, he says, no good deed goes unpunished, and irradiation might allow producers to "speed up the kill floor and spray shit everywhere."[51] A former editor of *Meat and Poultry* magazine agrees, arguing that irradiation will allow the meat industry to continue its unsanitary methods of producing meat. He says, "I don't want to be served irradiated feces along with my meat."[52]

Artificial flavorings, ubiquitous in American food, allow a direct comparison to genetically modified food because both are allowed under the FDA's GRAS (Generally Regarded As Safe) system. Under GRAS, if a tomato has an extra few genes that create a protein identical, or very similar, to known proteins, and if such proteins create no known food intolerances or allergies, the new tomato is considered safe. The same system allows artificial food flavorings to be judged safe, flavorings that Americans have been consuming for many years.

So what's in those artificial food flavorings? They consist of minute amounts of very potent chemicals, created after years of study and made in chemical plants. Artificial strawberry flavor consists of:

> amyl acetate, amyl butyrate, amyl valerate, anethool, anisyl formate, benzyl acetate, benzyl isobutyrate, benzyl acid, butyric acid, cinnamyl isobutyrate, cinnamyl valerate, cognac essential oil, diacetyl, dipropyl ketone, ethyl butyrate, ethyl cinnamate, ethyl heptanoate, ethyl heptylate, ethyl lactate, ethyl methylphenylglycidate, ethyl nitrate, ethyl propionate, ethyl valerate, heliotropin, hydroxyphrenyl-2-butanone (10 percent solution in alcohol), α-ionone, isobutyl anthranilate, isobutyl butyrate, lemon essential oil, maltol, 4-methylacetophenone, methyl anthranilate, methyl benzoate, methyl cinnamate, methyl heptine carbonate, methyl naphthyl ketone, methyl salicylate, mint essential oil, neroli essential oil, nerolin, neryl isobutyrate, orris butter, phenethyl alcohol, rose, rum ether, γ-undecalactone, vanillin, and solvent.[53]

For the average person who eats cookies or breakfast tarts with artificial flavoring, and who trusts scientists and federal agencies to insure that such ingredients are safe, isn't it contradictory to distrust the same scientists and government officials when it comes to genetically modified vegetables?

CONCLUSIONS ABOUT THE SAFETY OF MEAT

Although this book is not an exposé of the meat industry, we must have some base of comparison if critics are going to say that a tomato with a few new genes out of thirty thousand might be unsafe. Compared to the dangers we face daily from eating beef, pork, and chicken, slightly changed vegetables seem hardly dangerous at all. The inherent dangers in our large, almost global, meat system create many risks from eating meat and, next to such risks, eating genetically modified plants seems just plain safe.

This section has emphasized the risks of meat and not put them in context. In fact, the chances that the average American will get sick from eating meat are small and, overall, the system is fairly safe. My point is not to criticize the meat system but to compare acceptable risks there with acceptable risks from modifying apples and tomatoes.

Thus, I conclude that the risks from genetically modified fruits and vegetables are miniscule. Especially since these fruits and vegetables have been subjected to more thorough testing and monitoring than any new food in American history, they are very safe. No one has been harmed to date from eating genetically modified food, a claim that cannot be made about food produced by the meat industry.

Progressives will accept this view, but will everyone else? Globalists delight in a McWorld, for that is their vision, but egalitarians do not. The latter would point out that the same desire for maximal profits that created unsafe practices in the meat industry could create unsafe practices in agricultural industries. It might be just a matter of time before safety in genetic agriculture clashes with maximal profitability.

Naturalists also emphasize that two wrongs don't make a right. Although they would agree wholeheartedly that our industrial meat system is unsafe, unhealthy, and immoral, they would not accept the conclusion that genetically modified plants are therefore acceptable. They would likely reject both systems, emphasizing that the same demand from the vast buying power of fast-food restaurants could create, say, a standardized potato or standard loaf of bread for the entire planet, grown in a half a dozen huge monotracts. Because they oppose McWorld, they would oppose the means to this end.

VOLUNTARY VERSUS INVOLUNTARY RISK TAKING

Shouldn't consumers volunteer for risks? Shouldn't food with GM products at least be labeled? Isn't that why informed consent started to be required in medicine, because prior to the 1960s some physicians abused research on humans and never informed them of the risks? Isn't stealth introduction of new genes into traditional American foods like ghost surgery by residents without the informed consent of the patient, who thinks the famous surgeon is doing the operation?

One of Consumers Union's directors, Edward Groth, says,

> One of the most important factors is whether risk is voluntary. Consumers may not know if it's a big risk or not. But they get angry and resentful if the risk is imposed on them, rather than freely chosen. We scientists can get overly fascinated with our quantitative risk assessment methodologies. I think the public has a lot of common sense in making judgments about risks based less on scientific estimates and more on values and ethical questions. Is it fair? Is it necessary? Is it avoidable? Is it something I control myself or is it something someone does to me?
>
> Experts sometimes dismiss such public perceptions of risks as "irrational" or "emotional." This is a serious mistake. Public policy disputes are concerned centrally with whether or not a given risk is acceptable. That value judgment depends much more on . . . outrage factors than it does on the size of the risk. In a policy context, that's reasonable. In fact, focusing only on the size of a risk is rather "irrational," for policy purposes.[54]

Groth's view seems reasonable at first, but with reflection and comparison, it falls apart. Outrage is an emotional response—as with what has also been called the "yuck" factor, in response, e.g., to human cloning—and may be quite irrational. In the past, most people were outraged by the end of slavery and of racial segregation, by women's suffrage, and by gay rights.

Consider two examples in which it is far more reasonable to be outraged. First, federal investigators recently revealed that no reporting system exists for most of the products sold in health food stores or in the offices of practitioners of "complimentary medicine."[55] Manufacturers do not need to report to the FDA any injury or illness caused by dietary supplements like vitamins, minerals, herbal concoctions, amino acids, and products claiming to aid athletes. In other words, and even though many of these products are taken as substitutes for prescription drugs (e.g., St. John's Wort for Prozac), no adverse incidents are reported to the FDA, as they are for prescription drugs.

This lack of oversight is very dangerous. Investigators for the Health and Human Services Department reported that ginkgo biloba "can lead to excessive bleeding and can cause stroke" and that high dosages of vitamin A can cause birth defects during pregnancy. Supplements containing Aristolochia (a.k.a. Virginia snakeroot) can ruin the kidneys and cause urinary cancer.

This comparison shows the danger of appealing to what above was called "the outrage factor." Consider a second example. More than 100 million Americans take dietary supplements, many of which are bought in health food stores or similar sections of grocery stores. Most of these Americans probably assume that some medical, scientific, or public health organization certifies such products as healthy, or certifies the claims that are made about such products, but that is not true. At the same time, some of these same customers of health food stores shun genetically modified food as unnatural, undoubtedly unaware of the extensive testing of GM food.

Another example of how the outrage factor about genetic veggies may be muted by reflection concerns the continuing practices of allowing the meat of dead animals to be fed to other animals that are then fed to humans. Have the lessons of vCJD not been learned? *Consumer Reports* correctly criticizes the FDA's current rule where "cattle remains can still be fed to other animals, such as pigs, whose remains can then be fed back to cows."[56] Even more outrageous, "It is legal today for a herd of scrapie-infected sheep, or deer or elk with chronic wasting disease, to be used as feed for hogs or poultry. Their remains can also be used for pet food." In the same month that this issue of *Consumer Reports* arrived, *New Scientist* reported that some researchers were shocked to discover "that one strain of scrapie causes the same brain damage in mice as sCJD," and that the old dictum was false that scrapie could not be transmitted to humans.[57]

On reflection, we should be outraged by the lack of regulatory oversight of the nearly four thousand dietary supplements sold by the alternative medicine stores and even more outraged by the continuation of suspicious feeding practices in the animal industry. The "yuck" factor, then, is relative to what you know.

Finally, if we're going to label GM food, why not label all food? Why not label when the apple was picked, when the fish was taken from the sea, and when the pineapple was canned? Why not label exactly how much fat is in a cut of beef? If labeling is to occur, let's be consistent and label all foods for the benefit of consumers.

PESTICIDES IN FOOD WITH NEW GENES

The term "pesticide" covers a group of chemicals designed to kill organisms harmful to food plants and divides into three kinds: insecticides (to kill insects), herbicides (to kill weeds), and fungicides (to kill fungi). As any backyard gardener knows, pests can quickly eat up the food one wants to eat. Worldwide, as much as a third of crops are lost to pests, especially in areas lacking chemical pesticides.

In America, the Environmental Protection Agency (EPA) regulates use of pesticides on crops designed for consumption as human food. Its basic measure for a pesticide is a reference dose (RFD), which is the maximal amount of RFD's that humans may safely be exposed to *over a lifetime* and, for added safety, *divides that figure by anywhere from ten to ten thousand.*

One glitch in the American regulatory process is that the EPA does not test food animals for pesticide residues or other toxins, as such regulation falls under the Department of Agriculture. The USDA is traditionally seen as much more friendly to farmers than the EPA is.

Some years ago, some people became alarmed when it was pointed out that a diet of fruits and vegetables might be more dangerous than a diet of meat and cheese because of the resulting exposure of vegetarians to residues of pesticides. Specifically, pesticides may harm children more than adults because children may ingest more (from playing on grass sprayed with pesticides or putting sprayed objects in their mouths) and because their immune system may be less protective. Ingestion of pesticides in children is thought to be a silent killer by inflicting children with leukemia and brain cancers.

Vegetarian parents worried especially about children being exposed to pesticides. In response to these alarms, the American Academy of Pediatrics judged that a diet rich in fruits and vegetables was still, despite the small residues of pesticides, the most healthful diet a child could have.[58]

It is also true that consumers themselves can reduce exposure to even the normal residues of pesticides on fruits and vegetables by first washing and then scrubbing them with a small rubber brush where there are pits, holes, dirt, spoiled spots, or other damaged areas.

Of interest for our topic, the desire by the EPA, consumers, and farmers (because of costs) to reduce use of pesticides led to a new philosophy during the last decade of integrated pest management (IPM). IPM does not attempt to destroy all pests, but instead accepts a certain amount of damage in return for not using pesticides (which makes sense financially, since the cost in loss of crops may be less than the cost of pesticides to save the crops). Pesticides in IPM are only used when pests or weeds reach threshold levels. IPM also uses "good bugs," such as ladybugs, to eat bad bugs.

One good bug is the previously mentioned bacterium, *Bacillus thuringiensis* (Bt). Adding the Bt gene to corn to fight already well-known pests to a food crop is a logical extension of IPM. Soybeans, cotton, corn, and potatoes with Bt inside may require less herbicides and hence should contain less residues of herbicides. In addition, reducing use of herbicides also leads to cleaner waterways and, indirectly, cleaner water for drinking.

With the pest-fighting genes inside the corn, less spray need be used, reducing damage to the environment. So the alleged dangers from pesticides in genetically modified crops may turn out in the end to be a victory for progressives, who can accept the premises of naturalists ("Crops for human consumption should contain less residues of chemical pesticides") and use it to argue for such GM crops as Bt corn and Roundup Ready soybeans.

The bottom line is that residues of pesticides from spraying chemicals are greater than those obtained from genetically modified plants that genetically ward off insects. If the main worry about safety is residues of pesticides, then GM food is safer than traditional food.

Finally, it is a common myth that organic foods are totally free of pesticides, whereas nonorganic foods are coated with pesticides. As we have mentioned, organic growers use Bt sprays as often as they like on their produce.

END COMMENT

Dangers resulting from transfers of genes to plants, which in turn create unknown proteins that will produce allergic reactions in humans, seem slight. We know the major proteins that create allergic reactions in humans and can test for them. Already the FDA, manufacturers, and food processors agree that food containing proteins that are known allergens must be clearly labeled as such. Indeed, such GM foods have not yet even been created, much less labeled and sold, and probably will not be.

Two processes in America work to insure little exposure of consumers to old or new allergens. First, food companies and the FDA, USDA, and other federal agencies test novel foods for allergenicity. Precisely because this process of review worked, the soybean with the gene for Brazil nuts was not brought to market. Naturalists cannot legitimately cite a success of regulation as a reason why regulation is insufficient.

Second, real dangers of creating allergens carry risks of lawsuits, at least in America. Allergic reactions can be severe, sometimes resulting in anaphylactic shock and death. What food manufacturer will voluntarily introduce any allergen into his food, especially by stealth, when potential victims could sue him? Given popular fears about the term "genetic" and its conjunction with food, this would be stupid.

North American consumers face much greater dangers from the slaughtering, processing, and distribution of meat, pork, and chicken. As recent events in England have shown with mad cow and hoof-and-mouth diseases, the meat system is far more vulnerable to disease.

Along the same lines, genetically purified agents such as rBGH and chymosin are safer than natural, bovine-derived substances because infectious microbes cannot easily contaminate the cloned line. In these cases, "genetically engineered" food is really "genetically purified" food.

Overall, compared to what we tolerate in eating meat from an impersonal, centralized, industrial system, the possible risks of eating plants with a few new genes are miniscule. Even the "tilt factor" argument, that our lives are so risky that we can't tolerate any more, seems unpersuasive because the risks are orders of magnitude higher for driving, playing contact sports, drinking alcohol, and (for many people, still) using tobacco. Moreover, the argument from synergy here is far more persuasive for meat (and likely disasters of infection in this system), than for genetic fruits.

The final episode of StarLink corn has unfolded as trucks, rail cars, and barges unload corn at the five hundred granaries in North America: each load is tested for StarLink corn and Cry9C.[59] Using what looks like a home test for pregnancy, inspectors randomly test kernels in each load, and if some turn red the load is rejected. In an international industry created to produce a standardized output, such testing is a gigantic headache, so much so that many farmers may not plant gene-altered beans or corn again. Checking the distribution of StarLink corn, like the problem of finding infected cows in England, illustrates what happens when a problem arises in our massive system, namely, that because the system is so vast and interconnected, any major problem puts the whole system at risk.

Because so much money is at stake and because the whole system can be disabled by a major problem, my second conclusion sides with egalitarians and naturalists. For it is about trust. Although no reason exists now to fear the safety of GM food, the question arises whether we should trust international agribusiness to alert us if evidence of dangers begins to mount.

The record of StarLink corn and TSEs in Europe gives reason for concern. There just is no profit in being canaries of impending doom. So strong regulation and testing by outside agencies will be needed to insure that new GM foods are safe, and that they continue to be so. For that new task, some new food agency is needed, like the FDA, that would combine the existing inspection functions of the USDA, FDA, and CDC, and insure that all foods are inspected and safe.

6

Genetically Modified Crops, Environmental Ethics, and Ecofascism

We recognize that separating humanity from nature, from the whole of life, leads to humankind's own destruction and to the death of nations. Only through a reintegration of humanity into the whole of nature can our people be made stronger. That is the fundamental point of the biological tasks of our age. Humankind alone is no longer the focus of thought, but rather life as a whole. . . . This striving toward connectedness with the totality of life, with nature itself, a nature into which we are born, this is the deepest meaning and the true essence of National Socialist thought.

—*Ernst Lehmann, professor of botany and supporter of Hitler's National Socialism, in* Biologisher Wille[1]

How has Greenpeace International arrived at the morally bankrupt position that preserving plants is more important than feeding millions of starving humans? What has gone terribly wrong in the ethical footing of this elitist organization that it has slipped to this terrible place?

The answer will be some time in coming, for we must go to the roots of a misguided movement today that in essence worships nature. I believe that philosophical theories of radical environmentalism lie behind most contemporary opposition to gene-altered crops. In particular, ecofeminism, deep ecology, and a mystical embrace of the inherent value of Nature lie behind the opposition of activist groups such as Greenpeace International to gene-altered crops anywhere in the world, even for feeding the starving. Countries such as Switzerland hire such theorists to advise them on national policies about farming and agriculture, and in the Netherlands, the government is now required to financially support Greenpeace.

Until recently, no one in practical politics paid much attention to theorists in environmental ethics, but their influence has been growing in recent years, in part because the shallow roots of their thinking usually go unchallenged and because it is so popular for everyone to be in favor of protecting the environment. (Papers in the field of environmental ethics typically exhibit the problem of preaching to the choir.) Over the last quarter century, these theorists have advocated more and more radical, more and more ethically meaningless stances in fields such as environmental studies and environmental sciences. The ideas behind modern environmental ethics itself have a theoretical pedigree, one which it is important to understand, lest history repeat itself.

I understand fascism as composed of at least the following three elements: (1) a few (or one) charismatic leaders who (2) rule dictatorially (often disguising their authoritarian nature) but who do not respect due process or democratic vote, and (3) an emphasis on emotion, passion, mysticism, and intuition—not reason, evidence, or analysis—as the bases of the values championed. Nazism and Mussolini's fascism fit these definitions and, as I shall argue, so do some modern environmental groups.

ROUSSEAU, THOREAU, AND EMERSON

Reverence for Nature and opposition to the artificial has a long intellectual history. To take an early example in the eighteenth century, Jean-Jacques Rousseau then argued that civilization and science were not uplifting—as the Enlightenment held—but corrupting.

As he writes in his *Confessions*, Rousseau had never believed this before, but seized upon the idea as a way of winning a prize for writing an essay that would be bestowed by the king of France. His subsequent themes—the noble savage, that only life on a farm in a Swiss canton would return humanity to moral and political purity, that children should be home schooled by doting parents, and that children should be left as wild and as free as possible—enabled him to criticize the very aristocrats who supported him during his life (in this they were like rich people today who love to support radical egalitarians whose positions contradict the values behind how the wealth of the donor was obtained).

Rousseau popularized many ideas that we still hear today: that a true intellectual doesn't care how he appears and may even dress and smell rather badly; that those with real intellectual heft must be critics of everything that most people assume to be good; that a good living can be earned by constantly writing and giving talks about how science, society, and technology only appear to be good but really are bad; that medicine and science go against God's original plan for us to live naturally, as our forefathers did, in a simpler, nobler life.[2] He was the worst kind of public intellectual: a poseur, a hypocrite—one who loved to tweak others but who could not stand to be tweaked himself; one who professed to love humanity but treated all those close to him shamelessly. All in all, he was a narcissist who would champion almost any theme if, first, it brought him fame and money and, second, it did not contradict themes he had previously endorsed (he was smart enough to figure out themes which satisfied both conditions).

Ralph Waldo Emerson was also a hypocrite of sorts; he seems to have venerated an idealized, romanticized nature as a kind of fleshed-out God; his fuzzy, supernatural, nineteenth-century transcendentalism influenced both the younger Thoreau and, much later, radical environmentalists such as John Muir, who came to a similar view, that God was in nature or identical with it.

Scholar Lawrence Buell correctly calls Henry David Thoreau "the first American environmentalist saint."[3] Although Ralph Waldo Emerson was both mentor and friend, Thoreau's differences with Emerson illustrate conflicting themes in early American beliefs about nature. For Emerson, nature is the supernatural, higher part of us, and we a part of it. Or, at least, nature as constructed from Emerson's well-appointed study in Concord or his leisurely walks on the best weather days made possible by the income from his wife's estate and by his own lecturing and writing about nature. Like the Mandarin bureaucrats of ancient China, who felt confined by the strict, traditional Confucian duties they carried out, and who fantasized about escaping from their cubicles into the wild mountains symbolized by Taoist landscape paintings on their office walls, so Emerson's nature was a *romantic construct* that served as a foil against which to condemn machines and urbanization. For example, Emerson typically writes,

> He who frequents these scenes, where Nature discloses her magnificence to silence and solitude, will have his mind occupied by trains of thought of a peculiarly solemn tone, which never interrupted the profligacy of libertines, the money-getting of the miser, or the glory-getting of the ambitious. In the depths of the forest, where the noon comes like twilight, on the cliff, in the cavern, and by the lonely lake, where sounds of man's mirth and of man's sorrow were never heard, where the squirrel inhabits and the voice of the bird echoes,—is a shrine which few visit in vain, an oracle which returns no ambiguous response. The pilgrim who retires here wonders how his heart could ever cleave so mightily to the world whose deafening tumult he has left behind. What are temples and towers to him? He has come to a sweeter and more desirable creation. When his eye reaches upward by the sides of the piled rocks to the grassy summit, he feels that the magnificence of man is quelled and subdued here. The very leaf under his foot, the little flowers that embroider his path, outdo the art, and outshine the glory of man. . . . Things assume their natural proportions, before distorted by prejudice. What, in this solitude, are the libraries of learning? The scholar and the peasant are alike in the view which Nature takes of them. The barriers of artificial distinction are broken down. Society's iron scepter of ceremony is dishonored here,—here in the footsteps of the invisible, in the bright ruins of original creation, over which the Morning stars sang together, and where, even now, they shed their sweetest light. Whatsoever beings

> watch over these inner chambers of Nature, they have not abandoned their charge. Nature never abandons her house, and each year its glorious tapestry is newly hung.[4]

Nature is here mystified, deified, and imbued with ultimate metaphysical value, shorn of tetanus, rape, rabies, and a then-normal mortality rate of 25 percent before adulthood. Such a selective view of Nature, highly constructed through a rosy prism, blocks out anything nasty, dirty, deathly, common, or ugly.

Emerson's account is implicitly elitist. Workers of his time toiled long hours each day, six days a week, and few had the time or means for extended trips into the depths of nature. The "pilgrim" who sojourns deep into the forest finds "solitude" and "silence" as he escapes from "libertines." In these depths, it may seem "lonely," but such loneliness earns great rewards when Nature speaks her wisdom, surpassing man's artificial "libraries of learning." He is alone, far from the maddening crowd.

I believe that such uncritical elitism is inculcated into people as a motivational tool, but unchecked, it develops into dangerous political views. After hiking all day in the Rockies as a Boy Scout and carrying a heavy backpack, I would sometimes reach a vista at the end of the day. The leader would exclaim, "What a view! What a great achievement. Not one person in a million will ever see what you're seeing now." While even my youthful naïvete doubted his math, I did believe that most people would never see the top of, say, Wyoming's Wind River Range. Moreover, through these experiences, I believed at the time that only people who earned seeing these vistas by hiking up a hard trail should be allowed to experience these views. Only by hard work could the view be appreciated.

I now believe that I was incorrect. I believe that elderly people unable to climb or disabled people could appreciate the view just as much as I, even though they may only see it from a roadside turnout. Moreover, I think a roadside campground could afford many more people the opportunity to enjoy fine vistas, and I cannot explain why their pleasure is worth more or less than mine.

In short, a puritanical moralism infiltrated my adolescent aesthetics, such that I *wanted* to believe that I was one of the elite few who both had seen great views in the Rockies and, equally important, who had done what was required to *deserve* to see such views. Because I believed that ordinary people did not *deserve* to see such views, I was a junior ecofascist.

I think there are similar, elitist feelings behind the "eco warriors" of Greenpeace, the tree spikers of Earth First!, and those of the Earth Liberation Front who recently committed arson on homes being built too close to areas they considered wilderness in Colorado, New York, and Oregon.[5] When Dave Foreman of Earth First! once came to my university, his emotional, elitist presentation received wild applause from the sympathetic young environmentalists whom he attracted. As I looked at the young, athletic, well-fed, white crowd, I though of Nietzsche's young golden lions.

Where Emerson wrote of Nature from inside the comfy glass panes of his winter study, Henry David Thoreau—fourteen years his junior and very influenced by him—increasingly lived by what he had written, breaking off ties from people and living hermetically at Walden Pond, on the Merrimack River, and in the Maine woods. Of course, Thoreau didn't just talk and write about nature, he lived it, famously.

Interestingly, it was just this action that led Emerson to break off their friendship. It was almost as if Emerson meant, "My God, Henry, I didn't really mean it! You shouldn't really *live* that way!" Or, as Emerson wrote cuttingly to Thoreau, in his failed, life-long effort to get Thoreau to be more ambitious in writing, lecturing, and gaining the respect of high society:

> My dear Henry,
>
> A frog was made to live in a swamp, but a man was not made to live in a swamp. Yours ever,
>
> R[6]

A smarting Thoreau came to see Emerson as a hypocrite, a man who writes of despising man, civilization, ambition, and dependence on others ("Self-Reliance"), but who acts as if he values wealth and income, a fine wife and home, ambition, worldly success, transatlantic voyages, the libraries and museums of London, conventional manners, and the respect of educated people. So Thoreau concludes, "But when I consider what my friend's [Emerson's] relations and acquaintances are—what his tastes & habits [are]—then the difference between us gets named. I see that all these friends & acquaintances & tastes & habits are indeed my friend's self."[7] Ironically, in living Emerson's ideas, Thoreau lost one of the only friendships he really cared about. Maybe they could have never been real friends because Thoreau was too fanatical, too obsessive, and too independent to be friends with such a complicated person as Emerson, who had a wife and children and a satisfying public life.

Overall, the writings of Rousseau and Emerson, and both the life and writings of Thoreau, exalt the idea of a simple life in Nature: it is the source of true wisdom, of finding the meaning of one's life, of finding true values—all by obtaining the metaphysical epiphany that appreciates the truth of the statement that Nature and the Transcendent Being are one. Hence, Nature has deep, intrinsic value, both inherently and indirectly as the pure well from which humans must drink to find moral truth.

Of course, if one rejects the implied supernatural premise, none of this makes sense. And traditional theists should be suspicious, because such mushiness justifies anything. Remember that in symbolic logic, any statement follows logically from a contradiction. So a mushy exaltation of the environment can justify both ecofeminism and sexism, elitism and democracy, atheism and theism, egoism and altruism, as well as capitalism, war, and sexual promiscuity.

Glorifying the environment and Nature does not always lead in good, liberal directions. As we shall see, such glorification is also associated with political fascism. Hitler and Nazi ideology celebrated the clean, high mountain air breathed by the wild, dangerous type of man (more on this soon).

For those who worship Nature, machines, science, and the complicated relations of a large, extended family are corrupting simulacrum that divert us from religious insight and true values. It is not accident that Nietzsche admired Emerson; Nietzsche's robust Zarathustra haunts the high, wild areas of the Alps (unlike Nietzsche, whose own health was so fragile that he became sick from the slightest cause). Hatred of civilization leads to Osama bin Laden.

The scholar Joseph Wood Krutch, writing his biography of Thoreau in 1948, actually writes that "we [live] in an age of growing complexity and despair," as if one caused the other![8] Both Krutch and Thoreau take as their symbol of creeping, evil civilization the danger of being hit and knocked over by something, be it a horse or a car. Yet who would voluntarily take a one-way trip in a time machine to Thoreau's 1848 or to Krutch's 1948? (Antibiotics were just becoming available in 1948 and there was no Internet and no personal computers). Wouldn't most people, if given the choice of a one-way ticket, prefer to go forward to 2048 or 2148, hoping that mortality had been conquered or at least that lives had been extended a few hundred years? Isn't it interesting that conservatives today look back to 1948 as a golden time, but Krutch sees his own time of one of "increasing despair?"

As Emerson learned, these ideas normally serve as a harmless foil for civilization or as an escape from overurbanization, but occasionally get out of control when fanatics take them seriously. Then, the original context is lost, and suddenly the claim that nature has intrinsic value is not just a starting position in debate from which one hopes, say, to secure a more favorable ruling from a local zoning board, but a position that now justifies hurting innocent humans. For every Thoreau, there is a Unibomber.

Indeed, people who experiment with living as Thoreau, i.e. as Noble Primitives, do not enjoy it for long (take this as a testimonial from a former Eagle Scout). As described in Jan Krakauer's best-selling *Into the Wild*, Christopher McCandless was a young man who, after graduating from Emory University, simplified his life by giving away his worldly goods to Oxfam, who tried to live even more purely than Thoreau, and who froze/starved to death a year or so later in an abandoned school bus in the winter in the Alaskan bush.

Thoreau, too, died young at forty-five, from a cold, and Thoreau's brother died long before, as a young man, from tetanus. In general, hiking and camping on vacation are one thing; trying to raise your own food, kill it, clean it, cook it, and preserve it for winter are quite another. What people enjoy is *imagining* themselves living purely and simply in nature, not the actual life.

To some readers, it may seem blasphemous to criticize Emerson and Thoreau, especially Americans who read these writers in school and who remember their singing praises of the outdoors. Undoubtedly, French readers feel the same about Rousseau. More generally, it may seem wrong, or helping the enemy, to criticize champions of the environment. Isn't this just what Big Business, out to pollute the environment with lethal toxins, wants?

But ecofascism consists in, among other things, just such an unwillingness to criticize values passionately held that control a culture. Critical thinking brings clean rays of sunshine through the fog. Over 70 percent of North Americans say that are pro-environment and want their political administrations to "do more to protect the environment." But what does that mean, left so vague?

One of the jobs of philosophy is to examine assumptions commonly not questioned. So we ask, "If the environment must be protected at all costs, does that mean we should ban cars? Increase the tax on gasoline ten times to reduce driving (and hence, smog)?"

I have been told it is taboo to criticize pro-environmental organizations such as Greenpeace and Earth First! because to do so will deprive them of funds and volunteers. This is exactly where I fear ecofascism, especially when I see the stands taken by the above groups. California for decades has been the most pro-environmental state, with a mindless opposition to any new power plants, so its current power crisis was predictable. A similar passion for low buildings and protected open spaces has made housing costs in the Bay Area and Colorado unaffordable to average people. My point is that we are only just beginning to understand the consequences when 74 percent of people agreed in a *New York Times* poll in the 1980s that, "Protecting the environment is so important that the requirements and standards cannot be too high, and continuing environmental improvements must be made regardless of cost."[9]

IS THERE OBJECTIVE VALUE OUTSIDE HUMANS?

There is an old debate in axiology, the study of value and valuing, about whether all value derives from the wants, needs, and interests of humans or whether some independent standard can ground such value. The nineteenth-century American pragmatist William James posed this question in a famous thought experiment.

> Conceive yourself, if possible, suddenly stripped of all the emotions with which your world now inspires you, and try to imagine it as it exists, purely by itself, without your favorable or unfavorable, hopeful or apprehensive comment. It will be almost impossible for you to realize such a condition of negativity and deadness. No one portion of the universe would then have importance beyond another; and the whole collection of its things and series of its events would be without significance, character, expression, or perspective. Whatever of value our respective worlds may appear imbued with are thus pure gifts of the spectator's mind.[10]

Kant agreed with James. For Kant, nature has no intrinsic value but only instrumental value in fulfilling humans purposes. Hence, there are not duties to nature and nothing in nature can have a "right" that humans are obligated to respect. Destruction of waterfalls or beautiful caves is wrong not because it is destroying something of inherent worth, but because the destruction deprives thousands, maybe millions, of other humans of the pleasures that others have had in experiencing such wonders.[11]

Kant was depressed by the dismal state of European philosophy that led to Schopenhauer's and Hegel's fuzzy mysticism (he would have been even more depressed had he lived to read Heidegger). He was correct in this reaction. Now we are at a state in the modern world where views in environmental ethics are no longer positions in philosophy books published by obscure university presses, but exert power in public policy. Radical environmentalists are burning down houses of people to return areas to the wild, or putting nails in trees to harm loggers. These actions can be justified only by a metaphysical view of Nature, a view which permits harming humans and their interests in the name of a higher Nature.

It is this same view that would deny starving people genetically modified food at the altar of environmental purity.

Most naturalists believe that the world would still have value without humans. In particular, they think that waterfalls, ecosystems, herds of caribou in northern Canada, dolphins and whales in the ocean, and the majestic flights of bald eagles have value, and will still have value even if all humans disappear from the earth. Even if no human benefited in any way from the existence of such animals, they say it would be better that these things existed than if they did not.

This is because, some naturalists say, there is more than one animal whose interests count on the planet. All sentient nonhuman animals have value, even if no humans exist to interact with them. To argue that only humans have value, and that animals only have value because of their value to humans, is speciesist. (Which among some people, means the argument is over, but more on that later.)

This general view of the intrinsic value of nature should be regarded suspiciously. With James, one must ask whether, if no humans existed, nature in itself would have value. Put differently, if there comes a day when humans no longer exist, then on that day won't value cease to exist? Values are human created and originated by humans. No one who is not a human can understand what value is. Before humans came along, the world merely existed, without value. As David Hume wrote, nature "has no more regard to good above ill than heat above cold, or to drought above moisture, or to light above heavy."[12]

Naturalist and environmental theorist Holmes Rolston III believes that nature is not valuable because humans desire things from it, not valuable because humans appreciate it. Instead, he asserts, nature simply *has intrinsic value.* Why? How can amoebae and grasses have value without humans? He answers that the world would be poorer off without them and that these organisms have a good that can go better or worse *for them.*

In making this move, Rolston borrows from animal rights philosopher Tom Regan, who previously argued that animals have rights because they have interests, and they have interests because they have a life "that can go better or worse *for them.*"[13] That is, beings that have interests are not just biologically alive, but have something that they are striving for, and which can be thwarted.

This valuing by Regan is a conceptually slippery area. The concept of an interest in the law is very broad and often spreads from its root meaning covering persons to covering more abstract entities such as corporations, shareholders, and states. But if no humans existed, none of these derivative human institutions would have any interests. The meaning of "interest" fades as it strays from humans and their institutions, e.g., to lower animated beings such as shrimp and ants, and when applied to plants or oceans only has an analogical or metaphorical meaning. Where it strays too far, it has no meaning at all, e.g., the "interests" of Gaia.

We can also see this issue from the view of rights. Rights can be seen as two basic kinds—negative and positive—and earning one does not mean having the other. Rights of noninterference ("negative rights") are rights to be left alone, whereas "positive rights" entitle one to some service from others, such as medical care regardless of ability to pay. The latter is just the opposite of an entitlement to be left alone. Negative rights include the right to association (to be left undisturbed by government and police) and the right to personal privacy in one's home. Other positive rights include the right of normal citizens to clean water and a working environment free of violence.

Therefore, we cannot move from saying that any animal has a right not to be killed (because it has an interest in flourishing) to saying that it has a right to being fed or having an ecosystem provided for it. When we consider that humans also have interests, and that such interests will conflict with the interests of nonhuman animals, then it will be difficult to justify that a nonhuman animal has a positive right against humans, i.e., a claim against humans such that if they fail to act to satisfy the animal's (positive) right, they are bad persons.

Similarly, for Rolston to say that an amoeba or ant has intrinsic value because what happens to them *matters to them*, even assuming the premise to be true, makes an unjustified move from the assertion of "value to them" to "obligation to us." Plankton have value to shrimp, and perhaps shrimp should be left alone to pursue plankton, but no one has any obligation to provide plankton to shrimp. No particular person is bad who could provide anything to plankton and who does not.

But wait a minute. Why should we accept the original premise? That premise contains the suppressed but implicit idea that a being can have interests without having desires. That is the false move. Something can only "matter" to an organism if it has desires. A shrimp has no desires, unless we define "desire" circularly, saying that moving toward food shows desire. In short, can anything "matter" to an organism without desires? I don't think so.

On a quite different point, when people attribute intrinsic value to Nature, they are commonly thinking as tourists about hiking in the Tetons or some similar national park, or living in an area that has been carefully maintained to look natural, not of the more destructive part of nature, e.g., an F-5 tornado. Indeed, a position that Rolston attacks refutes the thesis of intrinsic (good) value of nature:

> An ecosystem contains only the thousandth part of creatures that sought to be but rather became seeds eaten, young fallen to prey or parasites or disease. The Darwinian revolution has revealed that the governing principle is survival in a world thrown forward (projected) in chaotic contest, with much randomness and waste besides. The wilderness teems with its kinds but is a vast graveyard with hundreds of species laid waste for one or two that survive. Four billion species have gone extinct; five million remain. Nature, lamented John Stuart Mill, is "an odious scene of violence."[14]

Rolston's reply to the above attack on his views? The weak response that we don't have to see it that way and, like the weak solutions to the problem of evil, that if we take a long enough view, everything will surely turn out to be good. At this point, we are in the realm of faith, not of fact or of rational argument.

When naturalists such as Holmes Rolston III claim that a world without humans has value, they smuggle in some conscious, human-like being whose admirability and interests confer value on the imagined landscape. A person thinking about such a world may covertly smuggle himself into the picture and imagine himself (contradictorily) as both not existing and as observing this world (say, from an airplane). Or God may be introduced as a source of valuing, where Nature and animals have value because their existence pleases God.

But such suspicious beings must be eliminated from the evaluational context, so the key question must be rephrased thus: "Assume no God, no imaginary observers of the planet, no conscious beings at all. Then does anything have value? If so, to whom and why?" Of course, asking "to whom and why" begs the question by assuming that value can only arrive *to* someone, and that is the point in dispute.

There are millions of stars in the universe, thousands of other galaxies. There are planets somewhat like ours, and many totally unlike ours. Do any of these planets have intrinsic value? If there are four other planets just like earth, or four billion, does it matter? It only matters if some consciousness exists to perceive it and if that being has desires.

Nevertheless, Holmes Rolston III's views influence people. Rolston is a thesis adviser at Colorado State University, where his students and students of others have produced a slew of theses on topics such as overcoming dualistic Cartesian metaphysics, reintegrating the self in the wider environmental world, and "Theory and Practice in Radical Environmental Activism."[15] Moreover, almost all the students writing such master's theses (at least, the ones listed on the web site) obtained (or already had) jobs as state officials charged with protecting the environment, lawyers practicing environmental law, or professors teaching environmental ethics. Is there any doubt that such views influence public policy today and thus promote the view that environments matter more than people? That preserving genetic diversity matters more than feeding starving people?

As we delve deeper into the morass of *nonanthropocentric* theories of valuing the environment, we get into a holier-than-thou competition among *ecocentric* or *biocentric* theorists about who is the most pure. All the latter theorists agree that nature has intrinsic or objective value independent of humans. Indeed, this is the quintessential theoretical question of environmental ethics, and such theorists all strive to be nonanthropocentric. All believe that ecosystems have intrinsic moral value independent of human wishes and that it would be wrong to destroy such systems to further human wishes.

Aldo Leopold, a former Wisconsin biology professor, champions "the land ethic," which goes beyond mere humans and defines "right as what furthers the interests of the 'biotic community.'"[16] Hence, it is conceivable that under certain circumstances in which humans were polluting, overpopulating, and wantonly destroying trees, that it would be right to kill humans to preserve the ecological system. Baird Callicott, a disciple of Leopold, champions environmental holism over individual human rights, a view that, if taken seriously, allows some persons to kill, maim, and steal from others in the name of protecting the environmental whole. As the title of a paper at a recent philosophical meeting put it, "Are human beings a cancer on the biosphere?"[17]

Of great interest to the topic of foods with genetic additions, holistic environmentalists emphasize that the natural has value, whereas the artificial does not. Following Rousseau, we should oppose the artificial and restore the world to its natural state. On this reasoning, we should not only oppose Bt corn but all modern corn, which has descended over many genetic crossbreedings from the short, stumpy maize that does not taste good, look so yellow, grow so big, or have so much nutrition inside.

Food plants that have had genes added to them for the benefit of humans are seen by such environmentalists as wrong on three counts: first, they see them as unnatural; second, they fear that these foods will be planted in monocultures and reduce indigenous diversity; and third, they fear that gene-altered plants will infect, or dominate, local varieties and hence destroy the natural habitat.

THE SCOPE OF MORAL THEORY

The problem contained in the previous discussion is an ancient one in ethical theory, although answers to it are more often bypassed or assumed than made explicit. The question concerns where the natural breaks occur in the expanding circles of our moral concern for others. More technically, what are the criteria for inclusion in the moral calculus? Legally, how do we decide whose interests get to come under the rule of law? For example, various legal cases have involved whether animals should be regarded as more than property under the law and as having protectable interests.

The natural breaks seem to be the self, the immediate family, the extended family, workers like me (e.g., fellow journalists), the community, the tribe, the nation, humans currently existing, living humans and future humans, both human and nonhuman mammals, all sentient creatures (including lobsters and fish), and then . . .

we go beyond sentience into environmental ethics. On this other side, the circles widen to include plants, streams, ecosystems, the planet, and finally, the galaxies and any beings who might or might not inhabit them.

Once you accept the reasoning that any entity must count that has a good "that can go better or worse for it," and once you allow some humans to speak for the inclusion of entities that cannot speak for themselves (be they comatose or plains of grass), you start to slide down a conceptual slippery slope on which it becomes difficult to draw a bright ethical line.

ECOFEMINISM

Within the last decade, one branch of feminism has embraced deep ecology and argued that what is wrong with traditional anthropocentric ethics is that it is really phallocentric ethics. As we have seen, ecofeminism is one species of naturalism.

The essential claim behind ecofeminism is that the same mindset that pillages the environment also lies behind sexism, also lies behind racism, speciesism, and class oppression. At bottom, white male patriarchy causes all these terrible global diseases, and if male, analytic values prevail, it will not be enough to return to a simple life with organic food, shared meaningful work, and homespun clothes. Instead, the intuitive, women-oriented "ethics of care" must dominate.

As we saw previously, Indian naturalist-egalitarian Vandana Shiva believes that progressives tend to be analytic, reductionistic, white males bent on asserting mastery, control, and domination over nature.[18] She likes Big Dichotomies: natural/artificial, free/enslaved, female/male, intuitive/analytic, holistic/reductionistic. As such, she exemplifies one branch of *ecofeminism* and its approach to issues such as genetically modified crops.

Carolyn Merchant's *Radical Ecology* extends Carol Gilligan's "ethics of care" to the environment. Education professor Carol Gilligan alleges that a certain kind of thinking in ethics is natural to women in a way it isn't for analysis-oriented, rights-claiming men.[19] In this analysis, Merchant also follows Boston University's infamous Mary Daley and her *Gyn-Ecology.*[20] Following Gilligan's lead, Merchant argues that ecofeminists can get us to change from dominating, earth-exploiting patriarchy to valuing peaceful, nature-loving, holistic relationships.

One wonders what Gilligan, Merchant, and Daly think of women who disagree with their views. Are such women "un-womanly?" Sub-woman? Perverts? If desires for sex and love are natural, and if people don't have such desires, we would see them as deficient, or even, as sociopaths. Are analytic women, who think nature valuable only for Kantian reasons, mental sociopaths? If so, it is hard to see what is feminist or profemale about such implications.

DEEP ECOLOGY

Deep ecologists think that traditional biocentric thinkers are too shallow and that a bolder metaphysical vision must ground environmental ethics. An oft-cited originator of this view, Arne Naess, argues that by identifying one's self with Nature, one sees the intrinsic value of Nature and achieves a higher, more valuable existence than occurs from the typical dualistic, alienated, individualistic stance of Western culture.[21] This view, of course, is recycled Emerson. Enlightenment proceeds in ever greater circles emanating from the original, isolated self: to lovers, friends, colleagues, neighbors, citizens, humanity, all sentient beings, all life, and finally, to the universe. So the enlightened yogi will not be "alienated from anything" and "Each new sort of identification corresponds to the widening of the self, and strengthens the urge to further widening." So the claim is that progressive levels of enlightenment bring a feeling of connection to a wider and wider range of beings, or at the end, to Being itself.

Paul W. Taylor, a lifelong professor of ethics at Brooklyn College in New York, takes all this one step further in arguing not just for the biocentric good of ecosystems but of the "inherent worth" of (here I list three kinds of organisms as examples) individual ants, slugs, and clumps of algae. Taylor argues that (1) humans are not the conquerors of ants, slugs, and clumps of algae; (2) humans are mutually dependent on organisms such as ants, slugs, and clumps of algae; (3) ants, slugs, and clumps of algae each have a way of flourishing or degrading; and (4) ants, slugs, and algae by their nature are morally coequal in worth with humans.[22] Therefore, he concludes, all life forms have equal inherent positive worth. Although this argument is valid, it is also unsound. Premises 1, 2, and 4 are easily rejected. Indeed, who but the most radical environmentalist would regard them as true? Yet premises are supposed to be the places where we all can begin to reason together.

I agree with commentators such as Louis Pojman, who observes that the implication of Taylor's view is "misanthropy: Killing a wildflower is tantamount to killing a human and may be worse" and, "From the point of view of the ecosystem, humans are unnecessary, gratuitous, spongers, and parasites." As Pojman and others observe, the ultimate reductio is that deep ecology implies that the most moral act for humans, in preserving the environment, is simply to eliminate the human species. If a virus was created, highly infectious but lethal only to humans and no other organism, deep ecologists might be morally bound to release it and to save the planet. But for whom?

DEEP ECOLOGY AND EASTERN MYSTICISM

Freya Mathews, a lecturer in philosophy at Australia's La Trobe University, thinks that identification with nature can dictate one's ethics. "Each viable self does its best," she writes, "within the terms of its own particular faculties, to further the interests both of itself and of the ecosystem through which it is defined."[23]

Similarly, William Devall and George Sessions champion deep ecology based on Eastern mysticism:

> The intuition behind biocentric equality is all things in the biosphere have an equal right to live and blossom and to reach their

> own individual forms of unfolding and self-realization with the larger Self-realization. The basic intuition is that all organisms and entities in the ecosphere, as parts of the interrelated whole, are equal in intrinsic worth.[24]

Again, what we have here is Emerson filtered through a selective Eastern mysticism (selected according to those parts the Western thinker wants, while he rejects other parts of the package, such as karma and caste).

The statement that all organisms in an ecosystem are equal in intrinsic worth is simply stupid. Any ethics that holds a human baby and an ant to be equal in intrinsic worth is also evil. Such an ethic would allow consistently evil results, for if both are equal, and since there are so many more ants than humans, then no destruction of any ant hill to build a house or to plant a field is ever justified. And why not sacrifice babies on the anthill? The utilitarian numbers justify it, when the greatest good includes all those beings of equal intrinsic worth. Why not just sacrifice humans to the multitudes, insects?

But of course, when ethics is built on Eastern/transcendental environmental mumbo-jumbo, people will project anything they want into the resulting view and call it justified.

Arthur Danto's classic work, *Mysticism and Morality*, warned us in the 1970s about those who scorned the methods of the West for grounding ethics and sought instead to ground them on the metaphysics of Eastern religions.[25] The metaphysical assumptions of such religions—atmans, karma, the wheel of rebirth, and nirvana—lack evidence and good arguments for their truth. But even if one accepted those assumptions, Danto taught us, Westerners could not get what they want for their ethics from them.

Like Adella in E. M. Forster's *Passage to India*, Westerners go to Asia seeking something they find lacking in Christianity or Judaism. But the lessons Adella learns in Hinduism are not what she expects. Eventually she is enlightened but not in the way she expected—the manner of the satisfying, even sexual, intoxication of St. Theresa on the couch receiving the Heavenly light from above—but through an unwanted insight that deeply disturbs her. Her new awareness brings turmoil, not peace; trauma, not serenity.

So with identification with the cosmos as a basis for environmental ethics. If I am one with the cosmos, nothing follows for ethics about how I should act or live because no matter which future state of the cosmos occurs, out of an infinite number of possible ones, I will still be one with it. It really doesn't matter when I am reincarnated as a Brahmin and in which life I die and then merge with Nirvana, for eventually I will, and then the drop of my atman will cease to exist and be absorbed by the oceanic self of Nirvana.

The irrelevance for ethics of self-identification with the environment can be seen in another way. On the street where I live, there are vacant lots on a beautiful river. Some people, even some of my neighbors whose names I know, dump trash on these lots, and the trash finds its way down to the river (if they didn't dump, they would have to pay someone to haul it away). Some of this trash is plastic, some contains car batteries, and there are also rubber tires.

How am I to achieve "oneness" with my neighbor who dumps his trash on the lots above the river? It would seem that I must obtain oneness with humanity before I can obtain it with wider phyla that include ants and wasps, but I stumble long before then. I believe that what he does is wrong, and I can't identify with him in any way.

The only way that I can achieve mystical peace about such dumping is to become indifferent, to reflect that he and I are both parts of the human system and the larger ecosystem, and that eventually the matter of both of us will be recycled and be reused by other organisms. My interaction with him to try to change his behavior will undoubtedly not change anything at all in the larger system. But this famous "freedom of indifference" of the mystics is a recipe for inaction, not environmental activism (it is also known in psychiatry as "la belle indifference").

On the other hand, the cold, analytic risk-benefit approach offers a pragmatic solution: require every citizen in the county to have garbage pickup and have bulk pickup one day a week or one day a month. Pay for it from taxes. Now people won't dump because they can get trash picked up on their front curb and because there is some risk associated with dumping (fines, neighbor's hostility). Score one for Western, analytic ethics. (A risk-benefit analysis might also argue for talking to my neighbor, but that is another calculation.)

But how am I supposed to derive the evaluative conclusion of radical activism from the premise of mystical unity? If I am one with Nature, or will be one day, what difference does it make what I do or don't do? What anyone else does or doesn't? What does or does not happen to Nature? In its Oneness, it is all the same One.

Nevertheless, radical environmental ethics somehow manages to derive not just liberal, democratic action out of such Eastern mysticism but radical *environmental* activism: activism that despises people and which has some strange implications.

For example, one champion of leaving Nature wild regards the domestication of animals as bad and immoral, saying that compared to their wild ancestors, domesticated horses and dogs are only "half alive." Sheep are characterized as "hooved locusts." Similarly, J. Baird Callicott calls these animals "living artifacts" that have been bred to "docility, . . . stupidity, and dependency."

Some critics observe that some animal liberationists seem to love animals and hate people. Paul Johnson's *Intellectuals* similarly describes a half a dozen famous intellectuals who seemed to love abstract humanity but mistreated all the people around them (Rousseau, Marx, Sartre, Tolstoy, Brecht).[26] Along the same line, some biocentrists seem to love ecosystems and hate people. Radical environmentalist Edward Abbey writes in his *Desert Solitaire* that he would rather shoot a man than a snake. Indeed, given the way some biocentrists rant about what evils humans are doing to the planet's ozone layer, air, oceans, streams, and land, one senses that the moral thing for humans to do would be to commit mass suicide. Humans thus will no longer be a cancer on the biosphere and Gaia can thrive without us.

RADICAL ENVIRONMENTAL THEORY AND FASCISM

The merger of biology and mysticism fused as worship of nature makes environmental activism appealing to some, especially to the young who do not know history. What young environmentalists never know is that environmentalism does not always lead to respect for persons, for democracy, and for due process. Especially when the end of preserving Nature or species is said to justify the means, people and their interests can get hurt. Especially when reason is abandoned for intuition, dangerous political consequences can follow.

The best example of such dangers was in the worship of Nature in Nazi ideology. Nature represented purity of a higher sort. The Homeland was pure, natural, untainted by "pollution." Hitler was a vegetarian, as were some of his top advisers. Although guilt by association proves nothing, vegetarianism and organic farming were urged by some Nazi thinkers as national policy.

The heart of Nazi *ecofascism* emphasized Nature, natural farming, environmental purity, and the homeland in order to divert attention from, and indeed to override, concerns about rights of individuals, egalitarian reforms, due process, and especially the respect due to peoples of all ethnic groups and nations.

Naturalist Wendell Berry writes about how elitist environmentalists scorn trailers and dirt bikes in scenic areas and, indeed, how they scorn people intruding on the pure wilderness they have created: "The conservationist congratulates himself, on the one hand, for his awareness of the severity of human influence on the natural world. On the other hand, in his own contact with that world, he can think of nothing but to efface himself—to leave it *just* the way it is."[27] One of Berry's friends laments city folk who come to places such as Maine, buy up land, and then post "No Trespassing" signs, denying an ancient local tradition of free trespass for hunting and gathering berries. This same friend says, when it comes to laying aside land for more wilderness, "I don't care about the landscape if I am to be excluded from it."[28] When the environment becomes more important than people, values are unsound.

It is also the heart of all fascistic movements that power is concentrated in a few, even one, charismatic leaders who brook no tolerance for elections or input from the troops. ("Captain" Paul Watson, "pioneer of direct action environmentalism"—according to his brochure advertising his availability for lectures at U.S. colleges—founder of the radical Sea Shepherd group, does not travel to hear what others have to say, but to be greeted by adoring fans. Ditto, Dave Foreman of Earth First! The same holds for Indian environmental activist Vandana Shiva and England's premier antigenetic food biologist, Mae-Wan Ho.)

Mysticism usually brims with contradictions. Hence, it is no accident then or today that biological/ecological mysticism can justify eugenics and racism: Nazi eugenics was based on mystical, biological notions of racial purity; Garrett Hardin's "let 'em starve" conclusion seems based on equally false notions of ecological fragility and overuse (more on Hardin later).

In his essay, "Fascist Ecology: The 'Green Wing' of the Nazi Party and Its Historical Antecedents," Peter Staudenmaier documents many connections between mystical worship of Nature and Nazi ideology.[29] German philosopher and unrepentant Nazi party member Martin Heidegger criticized technological urbanization and praised Nature's purity. Michael Zimmerman, a theorist of deep ecology, praised Heidegger as a deep ecologist:

> Heidegger's critique of anthropocentric humanism, his call for humanity to learn to "let things be," his notion that humanity is involved in a "play" or "dance" with earth, sky, and gods, his meditation on the possibility of an authentic mode of "dwelling" on the earth, his complaint that industrial technology is laying waste to the earth, his emphasis on the importance of local place and "homeland," his claim that humanity should guard and preserve things, instead of dominating them—all these aspects of Heidegger's thought help to support the claim that he is a major deep ecological theorist.[30]

Heidegger's fusion of Nazi and environmental themes build on the work of a previous German philosopher, Ludwig Klages, whose *Man and Earth* pushed the same fusion, especially to members of the German Youth Movement. As Staudenmaier sums up Klages's work:

> *Man and Earth* anticipated just about all of the themes of the contemporary ecology movement. It decried the accelerating extinction of species, disturbance of global ecosystemic balance, deforestation, destruction of aboriginal peoples and of wild habitats, urban sprawl, and the increasing alienation of people from nature. In emphatic terms it disparaged Christianity, capitalism, economic utilitarianism, hyperconsumption, and the ideology of "progress." It even condemned the environmental destructiveness of rampant tourism and the slaughter of whales, and displayed a clear recognition of the planet as an ecological totality. All of this in 1913!

After their country's loss in World War I, German youth searched for meaning. The so-called *Wandervogel* ("wandering free spirits") seemed like the protestors in Seattle: a "hodge-podge of countercultural elements, blending neo-Romanticism, Eastern philosophies, nature mysticism, hostility to reason, and a strong communal impulse in a confused but not less ardent search for authentic, non-alienated social relations."[31] Later, this naïve group would turn from worshipping nature to worshipping a Führer who worshipped Nature.

Hitler's mystical adoration of Nature stemmed in part from the influence of nineteenth-century figure Ernst Arndt and his student, Wilhelm Riehl. They extolled the purity of the German people and the German land, railed against urbanism and racial mixing, and called for protection of wilderness. Their views can be seen a half century later in a brochure of 1923 that was used to recruit members for its environmental organization and that echoes themes in the previous quote from Emerson:

> In every German breast the German forest quivers with its caverns and ravines, crags and boulders, waters and winds, legends and fairy tales, with its songs and its melodies, and awakens a powerful yearning and a longing for home; in all German souls the German forest lives and weaves with its depth and breadth, its stillness and strength,

> its might and dignity, its riches and its beauty—it is the source of German inwardness, of the German soul, of German freedom. Therefore protect and care for the German forest for the sake of the elders and the youth, and join the new German "League for the Protection and Consecration of the German Forest."[32]

Hitler's cabinet officers carried out the Nazi environmental ideology. Rudolf Hess, a more strict vegetarian than Hitler and devotee of homeopathic medicines, pushed organic farming and land preservation, especially on the willing agricultural minister, Fritz Todt. Hess also pushed the "Blood and Soil" connection, which asserted a quasimystical connection between the German people and their land. *Blut und Boden* was popularized by German racial theorist Richard Darre, who spoke of Jews as "weeds."[33]

Member's of today's Green parties in Europe seem to view their environmentalism as separate from Nazi ideology. This is a dangerous mistake. The importance of Nature in Nazi ideology had real consequences: it led to breaking up estates and holdings across Germany to make organic farms.[34] Hitler's and Hess's vegetarianism followed a devotion to purity and a horror of pollution that paralleled their thinking about race, eugenics, and ultimately, their actions in the Holocaust.

Moreover, there is a seemingly willful ignorance there in which anything connected to genetics is condemned as evil eugenics, whereas anything connected to the environment is wonderful. So the Green Party leads the fight against GM food and use of stem cells in medical research. Such a "pick and choose" approach to history serves no one's interest well.

Indeed, the theme of some radical environmental theorists today that people are a cancer on the planet is not far from the claim then that non-German peoples were a cancer on the German *Heimat* (Homeland). The acceptance today by Greenpeace and Vandana Shiva of starvation for peoples of developing countries to preserve environmental purity over acceptance of genetic veggies is not far from the claim, then, that racial and environmental purity must triumph over the needs of poor, non-German citizens. Finally, the claim that the power of the State must forcibly return industrialized society to a simpler, romanticized, purer, higher, country lifestyle underlies many environmental groups today and can be traced back through the Nazis at least as far back as Rousseau.

A book that deserved more appreciation, Ronald Bailey's 1993 *Eco-Scam: The False Prophets of Ecological Apocalypse,* detailed how in the 1980s more than 450 national organizations, and thousands more at the local level, raised money for the environment. Sometimes, it seems that such organizations mainly pay for expensive positions in Washington, D.C., where they focus on media relations and lobbying. Bailey claims the National Resources Defense Council "pulled in more than $400 million from a contributor base of nearly four million in 1990. Four hundred million is ten times the amount of money that the Republican and Democratic parties together raised in 1990."[35] What is taboo to mention is that raising money "for the environment" is big business and finances a lot of nice jobs, often with very little outside control. (Especially in California and the West, criticizing environment groups is seen as supporting polluting businesses.)

When certain groups become off limits to criticism, when certain issues become sacred cows, and when values become mindlessly advocated no matter what the opportunity costs, we are in dangerous territory. We should not repeat history.

END COMMENT

Deep Ecology isn't deep. Outside the circle of true believers, it's shallow and like all passionate, shallow movements, therefore, dangerous. Dangerous because it allows humans to be sacrificed for the interests of rabbits, deer, shrimp, molds, and algae.

What environmental ethics has never directly confronted, despite the hundreds of books that have been written on it, are the hard but obvious questions: to save one human baby, isn't it justified to sacrifice a billion ants? (My apologies to E. O. Wilson, but yes, it is.) To save one human, isn't it justified to sacrifice a species of tree forever? (Again, given this forced choice, we should save the human.) To save a million humans from starving, isn't it justified to sacrifice *everything* in their environment? (Yes, because humans matter over environments.)

It was entirely predictable that young environmentalists would be led to attack genetic veggies. After all, you don't need to know any biology or genetics to be a champion of the environment. Given the growing influence of courses in environmental ethics and environmental studies, and the focus on activism (like that of some feminist and sociology classes), one could have predicted that their graduates and professors would charge forward against GM food.

When faced with the damaging consequences of teaching their views, philosophy professors sometimes retreat by saying, "Well, it's only philosophy." But sometimes views in applied ethics *do* matter, especially over decades and if they become a movement. Remember Marx? (And Marxist philosophy professor Guzman led Peru's Shining Path.)

Hence, ecoterrorism predictably occurred over a hundred times in the United Kingdom and dozens of times in North America. In deep ecology and today's environmental ethics, ecological purity justifies harming human interests. No matter that genetic veggies might be good for starving humans or just good for everyone. Once labeled as bad for the environment, they must go.

In a recent review in *Teaching Philosophy*, the reviewer wrote, "Two years ago I taught environmental ethics for the first time. Not knowing much about the subject, I hunkered down for a few weeks with a half dozen of the leading textbooks. I was amazed by what I found. Virtually all of the texts shared the same radical agenda."[36]

He speaketh Truth. That agenda, when taken outside of the classroom by those eager to act, is "direct environmental action," a euphemism for demonstrations in Seattle against world trade, spiking old redwoods with nails to hurt loggers, committing arson against houses built near formerly natural areas, and burning field trials of genetic veggies.

Radical environmentalist ethical theories of the last decades, which have a pedigree a century and a half old, give intellectual cover and motivation for the recent ecoterrorism against GM crops. Such "direct action" should be seen for what it is: ecofascism, which, if it sweeps up millions more behind it, could become really dangerous.

If one is liberal and pro-environmental, it is easy to spot the "hot button" issues that politicians push to win votes without giving evidence or argument. By opposing abortion, unwed mothers on welfare, gay rights, prayer in schools, and affirmative action, some politicians automatically gain favor with those who support the status quo. Liberals who oppose them complain, "The masses are so unthinking! Whenever politicians press these buttons, they are led like sheep."

But those who champion women's rights, animal rights, rights to choose about abortion, gun control, gay rights, and the environment also have their "hot buttons." When someone is said to be "for the environment," that shouldn't be accepted uncritically. Hard questions should be asked about which humans lose and which benefit by a proposal, and how nonhuman interests can be weighed against those of humans.

What all this has to do with GM food comes in the following chapter. For now, we have at least established that not all human interests must be sacrificed at the altar of the environment.

7

Why Genetically Enhanced Food Will Help End Starvation

> Is it not as this mouth should tear
> this hand
> for lifting food to it?
>
> —*Shakespeare,* King Lear, *3.4*

Most philosophers, theologians, and social scientists who've written about starvation believe that its solution involves transfers of some kind from people in developed countries to those in poor countries. These approaches illustrate the egalitarian approach to solving famine: transferring resources from the rich to the poor. Whether such a transfer occurs by concerned individuals, national policy, or revolutionary change, the solution to famine is conceptualized simply as *transferring* something—food, land, money—from "haves" to "have nots."

Naturalists sometimes piggyback on this kind of egalitarian argument, for example, in the bumper sticker, "Live simply so that others may simple live." If people in developed countries became vegetarians, or bicycled to work, or lived communally, resources could be freed up to feed the starving.

As well motivated as these thinkers and their approaches are, I believe this approach will not really help starving people. I write here as an apostate, for once I accepted the transfer theory. Now I believe that some of its factual assumptions are false.

In saying this, I disassociate myself from a whole school of thinkers, beginning with Malthus and leading up to Paul Ehrlich, Garrett Hardin, and Lester Brown (and his alarmist World Watch Institute), who claim that people who think starvation can be cured are factually mistaken. I differ from these new Malthusians because I think they merely *claim* to use arguments about fact to disguise their real ideological commitments. Even worse are those trained in biology who use their scientific credibility with the public to make unsupported claims about why famine is inevitable.

Famine and whether it can be ended is a profound topic. Perhaps a billion people are hungry now and perhaps two billion will be hungry in the twenty-first century (the number depends in part, of course, on what we do). I believe that GM foods, especially those grown using chemically intense fertilizers rather than organic methods, will eventually feed the starving.

In this chapter I argue why previous egalitarian and naturalist approaches were flawed, why the gloom-and-doom of the neo-Malthusians is equally incorrect, and why a progressive approach using GM crops will play a goodly part in a milestone in human history: eliminating mass starvation.

CLASSICAL MALTHUSIANS

Due to the inevitable growth of humanity, the famous eighteenth- and nineteenth-century writer Thomas Malthus predicted that famine would always be a fact of life on our planet. Called variously the first economist, the first sociologist, or the first population scientist, Malthus inferred the future from the facts he knew.

Malthus, his predecessor Adam Smith, and his contemporary David Ricardo were the first globalists. Dubbed the "Manchester School," they founded laissez-faire economics, which,

> held, basically, that there is a world autonomous and separable from government or politics. The economic world, in this view, is regulated

> within itself by certain "natural laws," such as the law of supply and demand or the law of diminishing returns. All persons should follow their own enlightened self-interest; each knows his own interest better than anyone else; and the sum total of individual interests will add up to the general welfare and liberty of all. Government should do as little as possible; it should confine itself to preserving security of life and property, providing reasonable laws and reliable courts, and so assuring the discharge of private contracts, debts, and obligations. Not only business, but education, charity, and personal matters generally should be left to private initiative. There should be no tariffs; free trade should reign everywhere, for the economic system is worldwide, unaffected by political barriers or national differences.[1]

Laissez-faire economics began with Adam Smith, a professor of moral philosophy at Glasgow University during the Scottish Enlightenment. Smith held that an "invisible hand" made individual and social interests the same. This benign international market helped everyone, leading humans ever onward and upward.

Malthus published his famous (1798) *Essay on the Principle of Population* as a corrective to Smith's optimism and that of other Enlightenment thinkers. Malthus argued that human growth would explode until it was limited by lack of food, war, plague, or natural disasters. Crises could not be averted by scientific planning and, indeed, were Nature's way of checking population. To try to thwart Nature was foolish, and only created more misery later.

Two centuries before, in 1601, England had passed its famous Poor Laws, which influenced English life until 1834. The Poor Laws fleshed out an English worldview then that combined fledgling economics with Christian culture: while nothing could be done to prevent poverty, Christian charity should still prevent people from starving.

Malthus shocked people by coldly arguing for revocation of the Poor Laws. While feeding the poor relieved immediate suffering, the poor then just bred more, so in later decades there would be more starving mouths to feed. Contemporary historian Thomas Carlyle aptly dubbed this new Malthusian field of "political economy" the "dismal science."

MODERN MALTHUSIANS

Malthusian thinking continues today. Past Malthusians have opposed creation of food stamps, welfare, Social Security, Medicare, Head Start, Medicaid, vaccinations, preventive medicine, and public health reforms. For Malthusians, all such measures just allow the poor to breed more poor and only postpone the inevitable, natural crisis. Malthus was the first survivalist, and had he lived in America in 2000, would probably have had a cabin in northern Idaho stocked with five years' worth of supplies.

Malthus's view would be famously revived and defended by biologist Paul Ehrlich, whose sensationalistic book in 1968, *The Population Bomb*, scared citizens of developing countries. "The battle to feed humanity is over,": Ehrlich wrote, "In the 1970s and 1980s, hundreds of millions of people will starve to death in spite of any crash programs embarked upon now." Ehrlich did little more than recycle Malthus's basic ideas, merely updating the numbers and countries cited and adding a biological veneer.

In 1968 and throughout the next two decades, Ehrlich's views were given added weight by population ecologist Garrett Hardin, longtime professor of biology at the University of California at Santa Barbara. Like Ehrlich's, Hardin's views influenced so many for so long because he presented his view as *biological facts*, not as an ideological worldview. Hardin started with his famous idea of the "tragedy of the commons," where lands owned in common in English towns were overgrazed to extinction because the self-interest of each farmer led him to graze the maximal number of cattle.[2] That this and related views of Hardin appeared first in the journal *Science* led to their air of being irrefutable, factual matters.

In 1972, a sensationalized book, *The Limits of Growth*, by something called "The Club of Rome," followed Ehrlich's book and used the then-august authority of a computer model to prove that continued growth of human populations would exhaust the planet's resources in a few decades. That book built upon other, existing trends at the time, such as suspicion of big corporations, a desire of people to lead a natural lifestyle by living on cooperative farms (such as Twin Oaks in Virginia, patterned after B. F. Skinner's *Walden Two*), and a new emphasis in biology on systems, dubbed "ecology," all of which had led to events such as the first Earth Day in 1970.

Hardin's bleak ideas were popularized in a 1969 book *Famine 1975!*, which argued that Western nations should adopt military triage ethics. Only countries that adopted population-control policies and changed to live on their existing supply of food should receive aid.[3]

In a follow-up piece in 1976, Hardin advocated "carrying capacity as an ethical concept" and analyzed famine as an inevitable consequence of an ecological system's population exceeding its food supply.[4] Again, this was merely Malthus dressed up in the factual-sounding language of ecology.

Hardin further argued, like Malthus, that not only was it not our duty to send food to starving people in India, but that it was *harmful*. Future generations of Indians would be *worse off* because we had sent food. To Hardin, philosophical ethicists such as Peter Singer were mushy do-gooders who should learn some facts about human biosystems. The writings of such ethicists for Hardin were a classic example of the famous adage that, "the road to hell is paved with good intentions." In the following passage, Hardin cites an alleged "basic theorem of ecology" that "we can never do merely one thing":

> Surely—we think—a well-fed India would be better off?
>
> Not so: ceteris paribus, they will ultimately be worse-off. Remember: "we can never do merely one thing." A generous gift of food would have not only nutritional consequences: it would also have political and economic consequences. The difficulty of distributing free food to a poor people is well known. Harbor, storage, and transport inadequacies result in great losses of grain to rats and fungi. Political corruption diverts food from those who need it most to those who are more powerful. More abundant supplies depress free market prices and discourage native farmers from growing food in subsequent years. Research into better ways

> of agriculture is also discouraged. Why look for better ways to grow food when there is food enough already?[5]

Notice that Hardin specifically opposes funding projects to figure out ways to grow more food. In later decades, he tried to persuade the Rockefeller Foundation not to fund such projects.

So Hardin also argued that it was wrong to give food to India, and that giving food to Nepal in the early 1960s created too many people there, leading to deforestation, flooding downriver in Bangladesh, and starvation in 1974. He thought China and India had equally bleak prospects in 1926 but, because of food aid, China's aid-less 900 million became better off than India's aid-fed 600 million.[6] Giving food to India, he said, led to riots and civil wars, and was not really motivated politically by compassion but by the greed of American agribusiness, which profited handsomely from food-aid programs.[7]

His true ideological colors emerged when he argued for closing the borders of the United States to immigration. Admitting that most Americans are "children of thieves," he nonetheless persuasively argued for seeing America as a "lifeboat" into which the starving millions wanted entry. If we let too many on board, the whole lifeboat sinks.

Lester Brown, head of the World Watch Institute, is another famous Malthusian and author of *Who Will Feed China?* (1995) and *Tough Choices* (1996). In both these and many other books, Brown forecasts famines from overpopulation and gives readers the impression that no kinds of relief, new foods, or compassion will avert the coming problems.

Over the past two decades, Brown's institute periodically issued reports about the unstoppable, alarming increases in world population, which the print and visual media have uncritically reported as fact. The media do this despite the fact that the Green Revolution in agriculture averted the predicted world famine.

NATURALISTS AND EGALITARIANS: AGRIBUSINESS CAUSES FAMINE

Naturalists and egalitarians don't believe the rosy predictions of progressives about how genetically enhanced food will end famine. Peter Rosett, director of the Institute for Food and Development Policy and coauthor of *World Hunger: Twelve Myths*, argues that starving people are hungry not because of high population density but because of inequality in food distribution.[8] Similarly, gene-modified food is neither the best nor the only way to feed starving people.

Rosset's view is quintessentially egalitarian: "The real problems are poverty and inequality. Too many people are too poor to buy the food that is available or lack land on which to grow it themselves." For Rosset, the real enemy is the international conglomerates that want to profit by feeding the hungry and selling them genetically modified (GM) food.

Organizations such as Rosset's have been articulating their views on genetically enhanced food in expensive, full-page ads in newspapers such as the *New York Times*. This coalition calls itself "The Turning Point Project." Here's how one of their ads read:

> About half the world's population still lives on the land, growing food for their families and communities. But according to global agribusiness corporations, small farmers are not "productive" enough to feed the world, so here's what the corporations are doing:
>
> (1) Moving small farmers off the land, consolidating the small farm, into big ones under corporate control.
> (2) Replacing people with energy intensive machines.
> (3) Eliminating *diversified farming*, where different food crops are mixed and rotated to keep soil fertile.
> (4) Substituting *monocultures*—single crops grown over thousands of acres—headed for *export* markets.
> (5) Sustaining these monocultures with massive use of pesticides and chemical fertilizers.

(6) Wherever possible, using genetically engineered crops.
(7) Rewriting the rules of the WTO and the International Monetary Fund accelerated all the above; the rules accordingly favor global corporations over small farmers. This, the corporations say, will save the hungry.

> In fact, there's evidence of some increased output for a few individual crops in some countries, partly from these methods. During the past few decades, total agricultural production in the world increased. But hunger rates for most of the world have also increased, far faster even than population growth. What's the problem here?[9]

According to this coalition, use of genetically enhanced crops by agribusiness, even if proved safe, will not help local communities grow the basic foods they need for their traditional way of life. Instead, agribusiness will discover which crops grow best in the ecological niche of a country and, of these, which produces the most profit. This could be coffee in Colombia, shrimp in Ecuador, sugarcane in the Philippines, or beef in Argentina. Such crops will not be grown for local consumption, but for export to those who can pay the most, probably citizens of advanced countries.

In this process, as big corporations buy up good land using capital from people who invest in their stock, people without good land are reduced to farming more and more marginal land. As food becomes harder to grow on such land, local produce dries up, creating hunger.

Naturalists also claim that agribusiness creates few jobs, so people lack work, and hence wages, with which to buy food. Agribusiness that uses chemicals, fertilizers, mechanical harvesters, and genetically enhanced crops loses the thousands of workers who traditionally weeded and harvested the crops. As this pattern increases, the ad says, "cultures, communities and livelihoods are destroyed. Eventually, the farmers' families flee to crowded urban slums. Families that had fed themselves become society's burden, while corporate farmers get rich from exports."

Chemical and machine-intensive methods used by agribusiness leave the soil depleted; genetically enhanced crops used in monocultures reduce genetic diversity in crops.

The text of Turning Point ads pushes the egalitarian view that starvation will not be ended by growing more food but by redistributing existing food:

> [A]mazingly, the world has enough food right now. The UN World Food Programme reports that the world is now producing sufficient food to feed every person on the planet. The problem is that too much of it winds up in wealthy countries; poor countries have no cash to buy it.

Turning Point's members, such as the David Suzuki Foundation, disparage the claim that genetically enhanced crops will help solve world hunger:

> *Does anyone believe that producing biotech plants with sterile plants has something to do with stopping hunger?* 1.4 billion farmers on the planet now save seeds for future use; they have done this for millennia. But biotech "terminator" seeds don't reproduce. So, farmers may have to buy new seeds each year from corporations. Can anyone think of something more cynical than this?
>
> Even before inventing this *suicide seed*, biotech companies were already appropriating traditional seeds that farmers had bred over centuries. The companies make tiny genetic alterations, then *patent* the seeds, forcing farmers to buy what they formerly saved for themselves. Millions of small farmers cannot afford to pay these fees.
>
> Here's the obvious conclusion: It is not the goal of corporate agriculture to feed the world. That's just their advertising theme. *The corporations' goal is to feed themselves.*

From this viewpoint, the best way to cure starvation is to use the latest technology not to foster huge plots owned by international companies but small plots that are intensively cultivated and locally managed. The goal is not international niches but local self-sufficiency. Following this strategy, UN agencies and charities should sponsor clean-water projects and self-sustaining, agricultural practices. Seen from this viewpoint, international companies selling local peoples supercrops with terminator genes represents the worst possibility.

THE TRANSFER MYTH

North Americans and Europeans think of food as a simple thing perhaps because the minimal food they need to live (a bag of rice a month, a few gallons of milk, some fruit and vegetables) can be obtained for such a small percentage of their average income. Moreover, because obtaining food is indisputably necessary for the continued satisfaction of all other needs, it is easy to jump to the conclusion that getting minimal food is a "right" that can be demanded against those who have food. Prima facie, such a right would seem to justify redistribution of food.

The big problem here is this assumption that food is a simple thing. More often than not in the world, food is not simple. In the many developing countries of the world where starvation is always possible, food is not just a simple need but rather all needs and everything. There, controlling food is controlling life. Since the time of the Roman Empire, he who controls the growing, storing, and distribution of food in poor countries controls almost everything. It follows that food cannot be simply transferred from Republican Nebraska farmer-businessmen to starving Ethiopians. Doing so could only be quickly done by violating the rights of many people along the way who control food as their property.

There is enough food to feed everyone, as there was in the Roman Empire, but so long as the rich and their army control food, or so long as Polish serfs must work long days to produce food they cannot eat, no solution to famine will work that imagines it to be one of a charitable transfer from haves to have-nots.

The writings of Nobel Prize–winning economist Amartya Sen emphasize that famine takes place within a locus of legalized power relationships that allow it to occur. As he writes, "Market forces can be seen as operating through a system of legal relations (ownership rights, contractual obligations, legal exchanges, etc.). The law stands between food availability and food entitlement. Starvation deaths can reflect legality with a vengeance."[10]

What Sen is saying is that, in much of the undeveloped world, having power over food in developing countries is like having power over their savings, house, land, and inheritance for citizens of developed countries. Therefore, just as developers or environmentalists can't go into an existing suburb and raze it for their own ends, so well-intentioned philanthropists can't go into a poor country and *change its laws about property to end the power of dominant groups over food.*

Egalitarians think starvation is not the result of too little food but of too few people owning too much. Libertarian globalists agree with the premise but disagree with the solution: egalitarians would take away property from the well off, either forcibly or through external incentives, whereas libertarians think existing property relations are sacrosanct. Globalists believe that famine will be solved when a few countries specialize in growing food they can grow best and then trading such food to other countries for maximal advantage. Colombia can trade coffee for soybeans from Brazil or beef from Argentina. In this way, the efficiency of the global market benefits everyone.

Personally, I've always thought libertarians sanctimonious about property. They're really not interested in whether Rockefeller or the latifundia got their property justly, or even legally, just whether they have it legally now. Rich white people in America got rich centuries ago by stealing land and exploiting the labor of Native Americans, black slaves, Chinese laborers, and many poor whites who sold themselves as temporary slaves (indentured servants) to come here.

At least Garrett Hardin is honest enough to admit this, although he then argues that it doesn't make any difference now.

I do think Libertarians are correct that political systems that try to forcibly redistribute property almost always do more harm than good. Attempts ultimately failed in Russia, China, Cuba, and Nicaragua to redistribute farmland by force. Such redistribution did not stop subsequent famines in Russia and China, and ultimately these planned, centralized, overcontrolled economies failed because they could not provide food and goods as efficiently as market capitalism. Any great revolution in land rights will be accompanied for decades by chaos, deaths, and instability. Only the most coldhearted utopians can feel confident that such revolutions ultimately create the greatest good for the greatest number.

Yet little doubt exists that, when it comes to famine and use of genetically modified crops, property redistribution is what egalitarians really want. Especially in banana republics where latifundia have controlled 90 percent of the good land for centuries, egalitarians hope for new Bolívars, Che Guevaras, Fidel Castros, and radical Catholic priests who will force land reform.

I think a wiser policy is the gradual reform of property rights, e.g., heavy taxation of the latifundia in South America, and an American foreign policy of growing a stable middle class in these developing countries to add workers and consumer demand to a growing national economy. Here globalists are on the right track.

Egalitarians also have part of the truth. In North America, the savage effects of market capitalism and globalism have been tempered by a compassionate state that mitigates the effects of original inequality. Part of what has grown the huge middle class in developed countries has been huge moves toward equality over centuries. The pendulum here swings back and forth with political currents, and recently a laissez-faire attitude to technology companies has produced many new millionaires. But that should not make us overlook several great reforms in North American society that provided the ground support for rising aspirations: Medicare, welfare, food stamps, Medicaid, a graduated income tax, and in Canada, universal health care (including subsidized long-term nursing home care). Each of these reforms took money away from those who had it and redistributed it to more needy people.

I also want to point out that we too easily forget the past opposition of naturalist groups to scientific improvements that benefited humanity. For example, I remember a group called "Science for the People" back in the 1970s that adamantly opposed using new technology to create the Green Revolution, seeing such uses as suspect and as ignoring the real problem of structural inequality in society. Had food scientists listened to this group, millions would have starved to death.

Finally, even land reform can only do so much. Wars first must be stopped and civil order must be restored so that markets for food can be reestablished. Ethiopia had a great famine in the mid-1980s in which one million of its nine million people died. Then another such famine killed another million in 2000. Although food scientists tried to teach Ethiopians how to increase yields and gave them new crops, they couldn't stop the wars between Ethiopians, Eritreans, and people in neighboring Sudan. A similar war between Hutus and Tutsis destroyed the food system of central Africa.

IS SMALL BEAUTIFUL? SUSTAINABLE? THE NATURALIST VIEW

There does seem to be truth in the Turning Point Project's claim that a trend to global agricultural business throws people off family farms and makes them move to big cities. That occurred in eighteenth-century England and millions in North America have left farms over the last century for suburbia and cities. If the most efficient way to produce food is for 10 percent of the world to be farmers rather than 90 percent, it is only Ludditism to posit the rural economy as the ideal.

One of these questions is a value question, not a factual one. The American National Family Farm Coalition favors a world of small family farms, and there are important values supported by that way of life: the whole family working together, passing land down through generations, keeping people close to the land (versus working in cubicles inside skyscrapers, talking on headphones to people one never meets), and not being dependent on imported food in times of war or crisis.

On the other hand, is there any inherent reason why a family farm is more valuable than a family clock-making business, restaurant, or tannery? If all business was family run, the world would probably be inefficient because family farms don't have economies of scale. Family businesses also experience many unheralded problems, especially when the founder dies and when the children are not all equally eager to work as successors, or when personal problems within the family interfere with business, e. g., not every son or daughter wants to live the farmer's life.

The Turning Point Project then is surely right: farms all over the world, agribusiness, monocultures, and enhanced crops may accelerate the demise of small, family-run farms; agribusiness will plant the most profitable crops on its land and sell them to the highest bidder, which will probably not be the local people; agribusiness will use genetically enhanced crops to maximize production. But there still needs to be an argument that all this is bad for most people of these countries in the long run. After all, we can't ask them to preserve a simple, primitive life just so we can gawk as ecotourists.

This is a more general point about values and public policy (a topic to which we will return in the later chapter on food and environmental ethics) about the kind of farm and countryside a culture wishes to maintain. Egalitarians see great value in a nation of small, independent family farms. Moreover, they distrust international agribusinesses that are developing genetically enhanced crops: Vandana Shiva, Ph.D. in physics, author of *Biopiracy,* and director of the Research Foundation for Science, Technology, and Ecology, believes that small farmers will be unable to use these new kind of crops. She believes that these farmers should use organic methods that will not harm their health or the environment. "Agricultural biotechnology," she writes, "will aggravate the crisis that is already going on in places such as India, where farmers are committing suicide because they can't pay off their debts. Biotechnology is not sustainable over the long term for these small, vulnerable farmers."[11] She wants India to be a land where 80 million cattle supply two-thirds of the power requirements of villages, as cow dung is burned as fuel. About half of cow dung produced in India is used to fertilize India's fields, and is spread either by women's hands or by wooden carts used as manure spreaders. This is a technology that was current in America in the eighteenth century.

Continuing this line of thought, the coordinator of BIOTHAI, a group promoting the rights of farmers and biodiversity, says that, "The motive of the TNCs [transnational companies] is not really to help humanity." Withoon Lianchamroon claims that the goal of TNC's "is profit. We can't hand over the future of our farmers to them like that. Small farmers can't achieve security when transnational corporations control these technologies and give away GE [genetically engineered] seeds like others give out food aid. It doesn't work."[12]

A flash point in this debate has been Golden Rice. In May 2000, the Swiss-based company Zeneca acquired commercial rights to this product. Developed by the Rockefeller Foundation, Golden Rice contains forms of vitamin A that humans can easily absorb. The gene for beta-carotene gives the rice its golden hue. One might think of this as rice with a bit of carrot seasoning inside.

This development, amazingly, has been criticized by naturalist and environmental groups. "Malnutrition is a problem of poverty, not technology," said Day-cha Spirit of Thailand's Alternative Agriculture Network. These groups say the way to get vitamin A to poor kids in developing countries is through a diverse bunch of native crops, such as red palm oil and green leafy vegetables.

"What is most incredible with the vitamin A deal is that in the name of 'helping the poor,' Zeneca has acquired exclusive commercial control over a technology that was developed through public funding," three Asian antibiotech groups complained. (Although the Rockefeller Foundation supported most of the research on Golden Rice, Switzerland and the European Union also contributed.) "Why is this money being used to push expensive genetic engineering solutions," they complain, "rather than supporting sustainable alternatives to deal with malnutrition in developing countries?"[13] When Zeneca and the patent holders of Golden Rice agreed to give this rice to the world for free, the move was criticized by naturalists and egalitarians as merely a PR stunt.

These critics say that Golden Rice is really all about agribusiness getting quick, worldwide acceptance of genetically enhanced crops and then (can you say "paranoid") controlling world agriculture. (Zeneca can't win either way, can it?) As a result, fields planted to test Golden Rice have been burned in India and antibiotech groups in the Philippines stopped field trials by getting a judge to issue an injunction.[14]

PROGRESSIVE/GLOBALIST REPLIES

It is not necessarily true that selling lots of food for export can't bring a country wealth and create a middle class. The hidden assumption that Turning Point makes is the Marxist one that more and more workers will have less and less, rather than more and more, in building a middle class. The question then is whether developing countries will become banana republics or thriving economies.

Progressives here may part ways with globalists. Progressives try to create sustainable local crops that can be sold. Generally, the markets for such crops will be local and regional. Obviously, if local farmers can make more money by planting single crops for export, they will do so, but that seems to happen rarely, and is not typical in India, Pakistan, Indonesia, the Philippines, or Malaysia.

What is more important than a single crop versus diversified crops is having *some* regular crops that grow well, that can be predictably grown over decades and improved, and that can reinforce a cash food economy. And it seems that genetically enhanced crops can play an important role here, especially in bringing better-adapted varieties of rice, wheat, and cereals, as well as crops that carry vitamins and vaccines.

Nor is it clear that corporate ownership of food-growing land normally means fewer workers. That has certainly not occurred in California, where itinerant migrant workers still harvest most crops. Nor is it clear why a corporation would deliberately deplete its soil by only sowing one crop rather than rotating. Given the cost of acquiring and clearing land, many American corporations plant seedlings after forests are cut, such that another crop can be harvested decades later. As we saw with chicken, the industry has become spread over thousands of small farms in the Southeast, not a few huge chicken farms.

THE GREEN REVOLUTION OF THE 1970S

The Malthusians were wrong, very wrong, and the story of why is best told by the story of the life of a little-recognized hero of the twentieth century, plant biologist Norman Borlaug, whose work enabled perhaps a billion humans to live during the past two decades, people who otherwise would have died of starvation and starvation-related diseases.

In this work, dirtying his hands in the fields of real farmers, Borlaug continued the tradition of famous humanitarians such as George Washington Carver, who did a great deal for malnourished people.

Borlaug grew up on a small farm in Iowa and lived through the Depression there as a teenager. He studied forestry at the University of Minnesota, then studied plant pathology, including wheat rust, an airborne menace that evolved quickly, making it a dire threat to poor farmers and hard to eradicate.

Several years later, the Rockefeller Foundation sent a team of four biologists to do fieldwork in Mexico on improving crop yields and livestock management. As a plant pathologist, Borlaug looked for ways to prevent crop diseases without reliance on expensive fertilizers or chemicals.

Wheat was one of Mexico's most important crops, but yields were far below U.S. levels because of periodic epidemics of wheat rust. Borlaug crossed several varieties of native Mexican wheat, creating varieties by the random combination of genes. The methods used required great exertions: Borlaug wrote, "This hit-or-miss process is time consuming and mind-warpingly tedious. There's only one chance in thousands of ever finding what you want, and actually no guarantee of success at all."[15]

Such hit-or-miss methods are precisely what have been now abandoned because of new techniques of inserting just a few genes rather than randomly combining thousands and waiting to see if any good results occur. Luck and randomness can be reduced by many orders of magnitude, allowing humanity to more carefully control what plants do to produce the food it needs.

Nonetheless, Borlaug developed dwarf wheat, credited with being the essential crop behind the so-called Green Revolution of the next decades. The story of how he did this was told by Robert Herdt in a talk on Borlaug's life at Texas A. & M. University in 1998:

> Success in breeding wheat means keeping ahead of the ever-evolving rust organism so production doesn't crash. Failure means disaster for farmers, nations, and even in an extreme case, the globe. Borlaug became convinced that only by making thousands of crosses from wheats gathered from all around the world would it be possible to stay ahead of the threat. So he undertook to make many more crosses than any breeder had thought possible up till that time. That meant a tremendous increase in the fieldwork of examining and scoring the progeny. It also meant imposing tougher criteria for discarding plants that failed to measure up. This approach, high volume crossing, was the result of deliberate choice, and a willingness to take on the work implied by that choice.
>
> The second breeding innovation was more a matter of serendipity—one of those consequences that springs entirely unforeseen from innocent acts. The wheat program, like the other parts of the Foundation's effort, was being conducted in Toluca, not far from

> Mexico City, an area where most farmers were exceedingly poor. It wasn't particularly well suited for wheat production, and Borlaug's attention soon focused on the Yaqui Valley, 1200 miles to the north, in Sonora. The wide, flat Sonora plain, with irrigation from mountain waters, promised a much greater potential for feeding Mexico. Despite opposition from several quarters, including the Foundation, Borlaug persisted and gained approval to extend the wheat work to Sonora.
>
> Growing conditions in Sonora were a dramatic contrast with those in Toluca. Much further North and in a near desert, as harvest time neared and temperatures soared, humidity dropped and winds were often strong. Most wheats that had performed well in Toluca just didn't hold up in Sonora. But Borlaug saw how to use the climate differences to advantage. They would plant their segregating materials in Toluca in May, harvest and then plant that season's selections in Sonora the next October. That way they would be able to advance the generations twice as fast. But it went against one of the dominant plant breeding dogmas of the day—which was that plants had to be designed for the particular environments for which they were intended.
>
> Borlaug's objective was simply to make the breeding process faster. The segregating populations were shuttled back and forth over ten degrees of latitude and from near sea level to over eight thousand feet of altitude. They were exposed to different diseases, different soils, different climates and different daylights: to winter in Sonora and summer in Toluca. The result was much more than simply a speeding of the breeding process. The plants that survived and performed well were well adapted to a wide range of conditions. The concept of shuttle breeding is now recognized as a way to achieve that kind of wide adaptability.[16]

Borlaug would later consult in developing countries around the world, but came back to Mexico in the 1950s at the invitation of Mexico's president to found the Centro International de Mejoramiento de Maiz y Triego (International Maize and Wheat Improvement Center), an institute devoted to creating high-yield wheats suited for developing countries. A decade later, his semi-dwarf Mexican wheat, which had been field tested around the world by Borlaug's trainees, was growing in India and all over the world.

Admirers divide Borlaug's life into four phases, the first of which was the above-described Mexican phase. This was followed by the India-Pakistan phase, in which he did the work that would earn him the Nobel Prize in 1970. Gregg Easterbrook, in an admiring article on Borlaug in *The Atlantic Monthly* magazine in 1997, wrote about this phase:

> Owing to wartime emergency, Borlaug was given the go-ahead to circumvent the parastatals. "Within a few hours of that decision I had all the seed contracts signed and a much larger planting effort in place," he says. "If it hadn't been for the war, I might never have been given true freedom to test these ideas." The next harvest "was beautiful, a 98 percent improvement." By 1968 Pakistan was self-sufficient in wheat production. India required only a few years longer. Paul Ehrlich had written in *The Population Bomb* (1968) that it was "a fantasy" that India would "ever" feed itself. By 1974 India was self-sufficient in the production of all cereals. Pakistan progressed from harvesting 3.4 million tons of wheat annually when Borlaug arrived to around 18 million today, India from 11 million tons to 60 million. In both nations food production since the 1960s has increased faster than the rate of population growth. Briefly in the mid-1980s India even entered the world export market for grains. Borlaug's majestic accomplishment came to be labeled the Green Revolution.[17]

Despite these achievements and Borlaug's receipt of the Nobel Prize in 1970, Garrett Hardin argued in 1976 that Borlaug's ideas ran against factual ecology: "Whether or not the Green Revolution can increase food supply as much as its champions claim is a debatable but possibly irrelevant point. Those who support this well-intended humanitarian effort should first consider some of the fundamentals of human ecology."[18] (Remember that for Hardin, "well-intended" and "humanitarian" are pejorative terms.) Here

Hardin's ideology clearly emerges: he was convinced a priori that not only *giving* food but also that helping the starving *grow* their own food would result in exploding populations. In retrospect, Hardin's views were about as factual as those of eugenicists of the early twentieth century who perverted the fledgling field of genetics according to their racist views.

Beyond the crossbreeding of new strains of wheat and cereal, several other important ideas sparked the Green Revolution. First, farmers learned high-yield farming techniques to grow old crops on less land. Primitive slash-and-burn techniques require more and more land. New high-yield crops reduced land in cultivation and saved forests from being cut down.

Second, primitive methods of farming, such as those common today in Africa, require lots of people, which usually means lots of children to help weed, sow, and harvest food. More efficient methods require less help. High-yield, good-paying agriculture requires an educated child. Over time, parents invest in fewer children but educate the ones they have rather than producing many uneducated children.

Third, inorganic fertilizers used in high-yield farming do not pollute as badly as slash-and-burn techniques. Moreover, such fertilizers should not be confused with inorganic pesticides. Inorganic (chemical) fertilizers for high-yield crops in Africa are "sustainable," to use a key word in the literature, and can stabilize food production for decades such that famine is averted.

Fourth, environmental groups such as Greenpeace International oppose high-yield farming to end famine relief. European Green parties prevented their governments from supplying chemical fertilizers to Africa.[19] Similar green groups, according to Norman Borlaug (and who would you rather believe? A man who's saved a billion people or elitists groups that have mastered media-savvy techniques and who side with Jeremy Rifkin?) pressured the World Bank and the Rockefeller Foundation to cease funding his high-yield food projects in Africa.[20] For Malthusian reasons, Garrett Hardin also urged the Rockefeller Foundation not to fund better crops for the starving.

High-tech farming methods caused environmentalists to go to war in public relations with Borlaug. Easterbrook writes,

> Environmentalists continued to say that chemical fertilizer caused an ecological calamity in Africa. Opponents of high-yield agriculture "took the numbers for water pollution caused by fertilizer runoff in the United States and applied them to Africa, which is totally fallacious," says David Seckler [director of the International Irrigation Management Institute]. "Chemical-fertilizer use in Africa is so tiny you could increase application for decades before causing the environmental side effects we see here. Meanwhile, Africa is ruining its wildlife habitat with slash-and-burn farming, which many commentators romanticize because it is indigenous." Borlaug found that some foundation managers and World Bank officials had become hopelessly confused regarding the distinction between pesticides and fertilizer. He says, "The opponents of high-yield for Africa were speaking of the two as if they were the same because they're both made from chemicals, when the scales of toxicity are vastly different. Fertilizer only replaces substances naturally present in the soils anyway."

According to Robert Herdt, Borlaug's legacy is identical to the Green Revolution: "high-yielding, well-adapted varieties created by plant breeding, appropriate fertilizer, and stable, remunerative prices to farmers. In one country after another, Norman Borlaug has helped governments to see that these elements are the keys to increasing food production."[21]

GENETICALLY ENHANCED CROPS AND STOPPING FAMINE

What then about genetically enhanced crops? Will they allow us to end famine? Norman Borlaug answers, "With the technology that we now have available, and with the research information that's in the pipeline and in the process of being finalized to move into production, we have the know-how to produce the food that will be needed to feed the population of 8.3 billion people that will exist in the world in 2025." [22]

Moreover, we may be able to do so without destroying the environment. Borlaug continues, "Modern agriculture saves a lot of land for nature, for wildlife habitat, for flood control, for erosion control, for forest production." He predicts that, "we will be able to produce enough food in 2025 without expanding the area under cultivation very much and without having to move into semi-arid or forested mountainous topographies." He concludes, "those . . . values [will remain] that are important to society in general, and especially to the privileged who have a chance to spend a lot of long vacations out looking at nature."

Given the projected increase of the world's population to between 9 and 12 billion over the next fifty years, there seems to be no realistic alternative to Borlaug's ideas. Otherwise, there is not enough land to be used by primitive farming methods to grow the food needed. But to use such land, ideological battles will need to be fought with naturalists who do not want high-yield techniques used in developing countries.

In an essay in *Time* magazine in 2000, Microsoft founder Bill Gates argued that only genetically enhanced food could feed the world and also cure its poor people of diseases caused by malnourishment:

> The U.N. estimates that nearly 800 million people around the world are undernourished. The effects are devastating. About 400 million women of childbearing age are iron deficient, which means their babies are exposed to various birth defects. As many as 100 million children suffer from vitamin A deficiency, a leading cause of blindness. Tens of millions of people suffer from other major ailments and nutritional deficiencies caused by lack of food.
>
> How can biotech help? Biotechnologists have developed genetically modified rice [Golden Rice] that is fortified with beta-carotene—which the body converts into vitamin A—and additional

> iron, and they are working on other kinds of nutritionally improved crops. Biotech can also improve farming productivity in places where food shortages are caused by crop damage attributable to pests, drought, poor soil and crop viruses, bacteria or fungi.[23]

Shortly after Gates's essay appeared, scientists announced the creation of another new strain of genetically enhanced rice that could boost yields up to 35 percent. This is the kind of Borlaug-type technique that has worked so well before. Already the rice has been field-tested in China, Korea, and Chile. One rice expert said this discovery could be very significant because almost half the world's population depends on rice.

On the other hand, a spokesman for Friends of the Earth retorted that yields could also be increased by the same 35 percent by using "better farming methods." If that means "organic" methods, Borlaug has characterized such a claim as "ridiculous," and even if tried it would require cutting down millions of acres of forests to create the needed land.[24]

Imagine if, a century ago, naturalist organizations had publicly opposed the discovery of vitamins. One can imagine naturalists arguing that if God had meant people to have such things synthesized and taken as pills, He would have made them that way in the Garden of Eden. And many naturalists still oppose use of the pesticide DDT, but let's not forget that DDT wiped out malaria in most of the world.

Florence Wambugu, a plant geneticist and a native of a small village in Kenya, argues that whether Africa should use GM crops should be up to Africa, not Greenpeace or Europe.[25] She scorns the activities of Greenpeace to prevent introduction of GM crops in Africa: "Greenpeace is a $100 million company. To keep that budget you have to be doing something and doing it well. European people are having opinions forced on them through manipulation and half-truths about how dangerous this [GM] technology is." Moreover, "Some aid workers . . . are being pushed into an anti-GM position from their European office."

Wambugu says that organic farming methods won't stop famine in Africa: "In developed countries, food is getting cheaper because they use more and more technology, but in tropical Africa it is getting more expensive because it is all manually produced. People with a small salary spend almost all of it on food. If we can increase food productivity in rural areas it will bring the price of food down, and generate more money for investment to turn the wider economy around."

She thinks that use of Bt corn in Africa will create "millions of tons more grain." Her own work has been on the sweet potato, a major food crop in Africa. Monsanto has donated the patent on its genetically enhanced sweet potato to all of Africa via the Kenyan Agricultural Research Institute, headed by Wambugu. The new sweet potato is resistant to a virus that periodically devastates this crop.

END COMMENT

In July 2000, seven academies of science urged rapid acceptance of high-yielding techniques to alleviate world hunger, including use of genetically enhanced beans, wheat, and rice.[26] They urged people *not to focus on the process of adding a desirable trait to an old crop, but on the actual effects of the new crop to people and environments.*

Right now it's important to grow food in local regions so inchoate food economies can build. Egalitarians are right: such peoples don't need monocultures for export, they need to eat. The above-mentioned academies agreed that biotech so far has only marginally aided small farmers in emerging countries and that such farmers should not have to buy back enhanced versions of their native crops.

But GM crops are tools that can be used for many different purposes. How a tool is used depends on the motives of the person using it: a hammer can build a house or kill someone. So with GM plants. Surely Norman Borlaug and his successors will use this tool to build food houses; surely critics are wrong that this tool itself will not help starving people.

8

Will Genetically Modified Crops Hurt the Environment?

Let me outline as briefly as I can what seem to me the characteristics of these opposite kinds of mind. I conceive a strip-miner to be a model exploiter, and as a model nurturer, I take the old-fashioned idea or ideal of a farmer. The exploiter is a specialist, an expert; the nurturer is not. The standard of the exploiter is efficiency; the standard of the nurturer is care. The exploiter's goal is money, profit; the nurturer's goal is health—his land's health, his own, his family's, his community's, his country's. Whereas the exploiter asks of a piece of land only how much and how quickly it can be made to produce, the nurturer asks a question that is much more complex and difficult: What is its carrying capacity? (That is: How much can be taken from it without diminishing it? What can it produce *dependably* for an indefinite time?) The exploiter wishes to earn as much as possible by as little work as possible; the nurturer expects, certainly, to have a decent living from his work, but his characteristic wish is to work as *well* as possible. The competence of the exploiter is in organization; that of the nurturer is in order—a human order, that is, that accommodates itself both to other order and to mystery. The exploiter typically serves an institution or organization; the nurturer serves land, household, community, place. The exploiter thinks in terms of numbers, quantities, "hard facts"; the nurturer in terms of character, condition, quality, kind.

—*Wendell Berry,* The Unsettling of America: Culture and Agriculture

A FABLE FOR OUR TIME: KILLER BEES

What worries environmentalists about genetically modified (GM) veggies is the fear that mutant plants will get out of control, like killer algae. The mongoose, introduced in India to kill cobras, now kills everything. Kudzu, introduced from Japan in the American South to control soil erosion, now kills shrubs and trees that it overwhelms. European starlings, introduced intentionally in New York City, spread everywhere in North America. Dutch elm disease, gypsy moths, and the Mediterranean fruit fly were unintentionally introduced and proved very harmful. This Accident Story is one of our most ancient stories, one that commands fear, the allegiance of our moral intuitions, and sometimes, uncritical acceptance.

But this accident fable can paralyze us if we make it an idol. Take the zebra mussel case previously described. Some ecologists act on the either/or assumption that all exotic species are bad, all native species are good. But native species growing out of control can be as harmful as exotics, and sometimes exotics that at first were harmful become beneficial. So the zebra mussel, a ferocious filter feeder, while clogging the intake pipes of water works, coal-burning power plants, and even nuclear reactors, also has filtered the waters of the Great Lakes of the excess nutrients and associated algae caused by runoff from agricultural and municipal wastes.

As philosopher Mark Sagoff writes in *Reason*, "Lake Erie, which had once been given up as dead by eutrophication, is now clear of the organic matter that had been choking it, wholly because of the mussel."[1] In turn, these mussels gather deposits at the bottom of lakes into breeding habitats for snails and other invertebrates that are the natural food of larger fish such as the yellow perch.[2] However, it is politically incorrect among most ecologists and environmentalists to credit this mussel with anything good, since it is a paradigm of a bad exotic, and hence anyone, such as Sagoff, who says different is highly criticized.

Similarly, it is anathema among some groups to admit that traditional crossbreeding techniques are more dangerous than the selective insertion of a few genes in GM plants. This is a shame, *because selective insertion of genes is an environmentally friendly and environmentally enhancing tool.*

Take killer bees. Using traditional crossbreeding, Brazilian beekeepers wanted to produce a bee that created more honey. They did so by crossing their native bees with an African bee that produced more honey. In so doing, they bred bees that randomly exchanged thousands of genes, hoping to get higher honey production but also implicitly accepting any other changes that might tag along.

As Steve Milloy writes,

> The cross-breeding worked, in that the hybrid bees produced more honey, but it failed miserably in that the native bees took on the aggressiveness of the African species, attacking people, overcoming native species and migrating up the South American continent into the United States. If the Brazilian apiarists could have employed biotechnology or single-gene insertion, they could have transferred only the specific gene that controls honey production to create mild-mannered bees that produce more honey. This example from the animal kingdom is a good way of illustrating what occurs in plant technology.[3]

Once again, after reflection, an intuitive category is reversed: at first, genetically modified plants initially appear unfriendly to a stable ecosystem, but in fact they promote stability as a more selective tool than random crossbreeding.

BT CORN AND THE MONARCH BUTTERFLY

As mentioned in chapter 1, organic farmers spray their crops with *Bacillus thuringiensis* ("Bt") a.k.a. "the natural insecticide." According to the University of Minnesota Agricultural Service,

> Bt is a naturally occurring soilborne bacterium that is found worldwide. A unique feature of this bacterium is its production of crystal-like proteins that selectively kill specific groups of insects. These crystal proteins (Cry proteins) are insect stomach poisons that must be eaten to kill the insect. Once eaten, an insect's own digestive enzymes activate the toxic form of the protein. The Cry proteins bind

> to specific "receptors" on the intestinal lining and rupture the cells. Insects stop feeding within two hours of a first bite and, if enough toxin is eaten, die within two or three days. For more than 30 years, various liquid and granular formulations of Bt have been used successfully against European corn borers and other insect pests on a variety of crops.[4]

For progressives, the message here is, first, that Bt has been used as a commercial pesticide for thirty years and, second, that it is so safe that organic farmers can use it and still call their crops "organic."

What is Bt used against? Bt is especially effective in corn against the European corn borer, a vicious, widespread pest against corn that also sets up a cycle of other pathogens, especially Furarium ear rot.[5] Of great importance, Bt corn does not harm or kill other organisms essential to integrated pest management, especially honeybees and ladybugs.

In 1998, John Losey, an assistant professor of entomology at Cornell University, caused a sensation when he and two associates published a short, preliminary note in *Nature* that 40 percent of the larvae of monarch butterflies died when fed Bt pollen.[6] This news was reported on the front page of the *New York Times* with a color photograph of a monarch butterfly.[7] This seemed to be the first evidence that genetically altered crops could endanger something in the environment that people really cared about.

Monarch butterflies spend winters in Mexico and fly north in two generations. The first batch lays eggs in May in milkweed in southern states, and then their progeny fly north to the Midwest, where they feed and lay eggs in early June, when corn is shedding pollen. Losey said that monarchs "may be in the right place at the right time to be exposed to Bt-corn pollen."[8]

Naturalists reacted with glee, saying that Losey's study proved "that Bt-corn pollen harms monarch butterflies" and incorrectly implying that, as a result, Bt-corn had been banned in Austria, France, and Germany (it had been banned because of fear of GM crops) and that Monsanto's Bt-potato division had been closed down by its new parent company, Pharmacia, because of similar evidence (the Bt-potato suffered when McDonald's, which buys 10 percent of American potatoes, bowed to fears about GM crops and announced it would not use them).[9]

Naturalists harped on the harm to monarch butterflies because these butterflies symbolized the pure, unspoiled America that everyone wants to visit in the National Parks. As a result of this publicity, the Environmental Defense Fund asked the Environmental Protection Agency (EPA) to limit the sale of Bt corn and to require planting of traditional corn around fields of Bt corn, which the EPA did.

GENE-ALTERED PLANTS AND BIODIVERSITY

Many naturalist and egalitarian writers take it as axiomatic that genetically modified plants will reduce the immense biodiversity of organisms passed down to us from the past. Marc Lappé and Britt Bailey think that the evil corporations that formerly gave us DDT, Agent Orange, and PCBs have retooled into biotechnology. As such, "Agricultural biotechnology threatens to decrease the number of crop plant varieties currently grown by substituting a few varieties for the many now in commerce."[10]

So economies of scale and desires for short-term profits will make companies sell, and farmers buy, only a few varieties of corn, soybeans, tomatoes, apples, and so on, resulting in huge crops subject to instant destruction when a new virus or predator arises. Crop uniformity contributed to the Irish potato famine of 1845, to the destruction of sugarcane in Indonesia by a virus in 1890, to "red rust cut" of American wheat in 1916 and 1954, to southern corn-leaf blight in 1970, to American citrus canker blight in 1984, and to the destruction of American wheat in 1989 by aphids and other insects.[11]

Potatoes came from the Andes, and the same blight that struck Ireland also struck the Andes, but because Andean natives cultivated many varieties of potato, some of these varieties had inherent resistance to the blight (these varieties were used to restore potatoes as a crop around the world). Point: keep lots of varieties of important crops lest your main crop gets wiped out and you have no backup.

We can agree to all the above and reject the conclusion that genes of plants should never be altered in new ways. Indeed, the ability to make new alterations gives us new weapons against such diseases. The underlying assumption of Lappé and Bailey is that farmers and corporations are too stupid to know their own long-term interest. Indian farmers have criticized the similar, simplistic rantings of Vandana Shiva, as seen in the reaction to her by the Indian farmer who writes:

> In India, for example, food self-sufficiency and higher life-expectancy has been achieved in the post-independence era, thanks to technology in spite of, rather than because of, government planning. Every technology creates its own environmental problems; these problems are resolved by a further advancement to higher technology and never by retreating into the primitive past.
>
> . . . Biodiversity is an important thing. Humanity cannot afford to lose any source of genes which may become one day an important source for introduction of a given trait. It is equally true that Mother Nature has been eliminating species every day and creating new ones all the time. It is not the job of farmers to preserve germplasms. That job must be left to specialists. Farmers will grow crops for which there is a demand in the market. If consumers demand plump red tomatoes, farmers will not grow indigenous varieties merely to please environmentalists.[12]

Farmers and governments will not put all their eggs in one monoculture basket.

The claim is absurd that GM technology will always decrease biodiversity. If I can add a million different genes to tomatoes, a few at a time and each producing a new variety, I have not reduced biodiversity in the world, but increased it. Plant and animal hybrids have been carefully bred for human advantage for centuries, but were limited by what could work in traditional methods of reproduction. The new technologies now allow us to "let a thousand flowers bloom."

Actually, critics of modifying genes of plants often contradict themselves about biodiversity in a number of ways. On the one hand, critics claim that people developing these plants will only plant monocultures selected to thrive in certain biological niches. Critics envision these monocultures as being so successful that they will engulf the ecosystem and kill all other plants. Perhaps they are thinking of kudzu or purple algae.

On the other hand, they worry that so many new kinds of plants and animal hybrids will be created that all sense of wildness and natural variety will be lost. Here, they may be thinking of horses, sheep, and dogs, and how certain breeds have been carefully selected for qualities valued by humans.

In each case, it is not the tool that is subject to criticism, but the end result of human decisions. Few naturalists, even the most radical, would want us to return to the ecosystem where only wolves and no dogs existed, but many would agree that enough breeds of dogs now exist and no new breeds should be created.

A very similar feeling exists about plants. The view is that, even if our present varieties of food crops, meadowland flowers, and trees in forests have developed after intentional and unintentional human interventions, ecosystems would be fine if humans just stopped suburban sprawl, overpopulation, polluting lifestyles, and unhealthy eating practices.

This is a version of "last one in" shuts the door, where the last Californian to leave a commuter city and move to a small town in Oregon votes to limit the very kind of growth that allowed him or her to move. Only here the view is that the species of plants and animals that have evolved to be here in, say, the year 2002 are just right and we should not let change occur again for a long time.

This view has two metaphysical underpinnings. One, of course, is that an omnipotent Brahman, Allah, God, or Tao guides and directs the ecosystem, producing homeostasis. This view ultimately contradicts itself, for if man's sinful plantings can thwart divine naturalism, then the deity is not omnipotent. If man's own changes are part of the divine plan, then *whatever* humans do is ordained: only what God allows will and can happen; humans cannot thwart his plan, even when they think they can.

A secular version of the Invisible Divine Hand View is also popular. Here evolution subs for God. Here the deep intuition is that "you don't mess with Mother Nature," especially after she has spent millions of years developing particular plants and animals uniquely adapted for both this planet and a particular niche in it. Behind that evolution and adaptation lay thousands of near misses, millions of organisms that coevolved in support, and all interlinked across ecosystems and the planet in mutual interdependence.

This is the objection of Stephen Palumbi, a professor of biology and curator of invertebrates at Harvard's Museum of Comparative Zoology who criticizes what he calls "brute-force" insertion of genes. Rather than allowing large sets of genes of organisms to evolve in ecosystems over millennia, brute force manipulation, he claims, is like dropping guns on Main Street and hoping things sort themselves out for a new practice area for a Girl Scout rifle team.[13] In other words, you get genes that express certain traits, but not necessarily the genes that regulate expression of those traits in different environments.

Palumbi also stresses that the plants and animals adapted to particular environmental niches survived after hundreds, maybe thousands, of similar variations failed. Ruthless competition among all these possible plants found the very best one for one specific ecosystem.

In contrast, humans with GM crops adapt a plant and by "brute force" insert new genes into it and then insert it into a new environment. As such, it has not passed evolutionary hurdles and, should a lethal virus come along or some old parasite reemerge, the crop will be wiped out.

Palumbi's critique boils down to a gut naturalist unease at GM techniques and two worries: crops such as Bt corn didn't survive evolutionary testing and Bt corn has no backup varieties against, say, a more potent corn-borer. To this, progressives reply that, while this is true of Bt corn, it is also true of *all* modern corn, which is far, far different from ancient maize, which originated in Mexico.

His point is well taken that if all farmers plant the same variety of corn, soybean, or cotton, all could be wiped out by a new virus or pest. But crop insurance can be purchased for this, other varieties can be planted as just such insurance, and corn in the rest of the world would then jump in value. Moreover, just for this reason, agricultural depositories now store germ plasm and seeds of thousands of plants.

As for Palumbi's point about evolution, what's so good about randomness, the key element in evolution? Could there be any more difficult and slow way to achieve progress than waiting for a random genetic mutation to give a species a competitive advantage during times of food or mate scarcity? This would be like sitting a million teenagers down at computer terminals and having them write for millions of hours in a misguided attempt to see if one could write a textbook of physics. Although the intended result is logically possible, it would be far easier to train some in physics, cull the best and send them to graduate school, and then let some of those with doctorates write the text.

In the same way, we can hunt around in the tropical rain forest for new drugs and insecticides, or we can create our own biologically diverse pharmacopoeia. There's no inherent reason why existing biodiversity in a tropical rain forest offers superior promise of life-saving drugs than ones created and tested just for this purpose in labs. Prying secrets out of shamans may create shortcuts and mine indigenous knowledge of plants that already cure, with potentially great benefits to humanity, but that is not the *only* route to that goal.

Indeed, as we learn more about ethnic and individual variations in responses to drugs (7 to 10 percent of the population can't properly metabolize perhaps 25 percent of drugs currently on the market),[14] knowledge of what existing plants do may become less important. What will ultimately unlock the secret will probably be the molecular basis of disease and cure, despite the naturalists' horror of such approaches.

I suspect that in biology a residual Invisible Divine Hand underlies the popular worship of biodiversity in biology. After all, a deism is compatible with evolution, at least if one allows the deity to work through evolutionary selection and not direct intervention. Something has to make randomness into something inherently good, other than the fortuitous fact that evolution has brought us to our present state.

GENE OUTCROSSING: HYBRIDIZATION

Some transgenic plants already help clean the environment of human-created toxins. In the first case of patenting of a transgenic organism, Chakrabarty famously modified a bacterium so that it would "eat" oil spills, such as that of the Exxon *Valdez* in 1989 at Prudhoe Bay, Alaska. Some plants have been created that screen out metals contained in metal-contaminated sludge that can be then used as fertilizer.[15] Still others create crops that are resistant to herbicides such as Roundup, so that less Roundup can be used, with subsequent benefits to the environment.

On the other hand, there are possible perils to the environment from transgenic plants. One major worry is about *outcrossing* or *hybridization.* Genes of the transgenic crops (for example, corn) might migrate to crops on organic farms or to wild varieties (maize).

As naturalist Marc Lappé writes, "A major concern is that the integrity of plant species will be compromised as engineered crop acreage increases and pollen mediated gene flow swamps and suppresses rare native plants that are congeners for transgenic crops. As remote as the present prospect for gene flow seems, herbicide targeted weedy species may pick up some transgenes and evolve more quickly than now anticipated."[16]

Jane Rissler and Margaret Mellon of the Union of Concerned Scientists, a naturalist-egalitarian organization, argue similarly:

> Scientists have little knowledge of the macroenvironmental consequences of producing transgenic plants. Given enough time and a broad enough selection of engineered crops, movement of transgenes into wild relatives of crop plants is a virtual certainty. Thus, the widespread adoption of genetically engineered crops will mean a constant flow of novel genes not only into agricultural ecosystems, but through such systems into wild ecosystems as well. Many of these genes will come from animals and bacteria and would never have found their way into plants without genetic engineering. What will it mean to have a steady stream of animal and microbial genes entering the gene pools of plants in wild ecosystems? It may mean little. But as yet, it is clear that we lack even a framework within which to answer the question.[17]

The problem with this prediction of danger is that it does not seem well thought out. The surrounding plants could only breed with new plants by traditional means of reproduction and a new gene introduced into, say, a tomato plant to increase resistance to frost is not going to transfer to a wild grass *because tomatoes and grasses do not interbreed.*

In a study in nature in Britain, some gene transfer of this sort did occur. Looking via satellite down at fifteen thousand square kilometers in southern England, researchers identified fields where traditional oilseed rape or canola was being grown.[18] Knowing that a wild turnip, *Brassica rapa*, grows along river banks, researchers identified two sites where the two relatives were likely to be growing next to each other, and hence susceptible to gene transfer via pollination. Testing leaf samples from 505 plants, they found one hybrid formed by cross-pollination. As *New Scientist* reported, "this suggested that hybridization between crops and weeds is rare—but does occur."[19]

Notice this tiny outcrossing occurred when a native plant grew right next to an altered variety. Steps could easily be taken to prevent this, for example, by requiring large buffer zones, as the EPA is doing now with Bt crops. But that assumes that the outcrossed crops would have some advantage in the surrounding environment, which may not be true and which would need to be judged on a case-by-case basis.

CONCEPTUAL ISSUE: WHAT IS A WEED?

Naturalists worry that modifications to traditional crops such as soybeans will allow these supercrops to grow advantageously outside intensively managed farms and become weeds in the surrounding habitat. According to Rissler and Mellon, crabgrass, that scourge of suburban lawns everywhere, was introduced in America to produce grain.[20] Similarly, Johnson grass, now regarded as one of the world's "Top 10" weeds by one evaluation, came to North America a hundred years ago as a grass for cattle and horses.

It is of interest here that the definition of "weed" is surprisingly slippery. Some concepts in biology are factual, some are quintessentially normative. Identifying a plant as a weed is identical to making a value judgment that the plant is undesirable for some human purpose or from some human point of view. There is no such thing as a factual classification of weeds and nonweeds.

For defining weeds, ethical relativism runs wild: a plant may be a weed to one person and a valued item to another. What the suburbanite sees as plant pests in his yard—fireweed, dandelion, St. John's wort, chickweed, oat straw, nettle—are seen as natural herbal drugs by practitioners of alternative medicine.

From a naturalist viewpoint emphasizing biodiversity, a weed is a plant that humans don't now know how to eat, make into a drug, or otherwise put to good use.

"Weed" may also be a legal term. The city of Santa Monica, California, defines weeds in its municipal code:

> *Chapter 7.44 WEEDS—Section 7.44.010 Definition of weeds; nuisance.* The word "weeds" as used in this section includes weeds which bear

> seeds of a downy or wingy nature, and any other brush or weeds which attain such growth as to become, when dry, a fire menace to adjacent improved property, weeds which are otherwise noxious or dangerous, poison oak and poison ivy when conditions of growth are such as to constitute a menace to the public health, and dry grass, stubble, brush or other flammable material which endangers the public safety by creating a fire hazard; and vegetation, vines and shrubs of every kind and nature overgrowing curblines or draping over walls or fences along, or projecting into, public streets, including alleys, thereby interfering with public street use and maintenance and the public safety; and all of the same are hereby declared to be a public nuisance. (Prior code §7700)[21]

Legally, if a plant is a weed, city employees may have the right to remove or eradicate it on private property—a right they may not have for other plants on private property.

Naturalists fear that transgenic crops may become weeds. Because the definition of weed is relative to a human purpose, such fears about old plants with some new genes will also be relative to specific human fears, such as the fear that new plants—now called "weeds"—may take over fields of farmers or national parks.

Like the shifting, evaluative concept of "weed," the concept of "toxin" is also relative and shifting. In bioethics, people often assume that it is easier and more factual to define harms than benefits, whereas in fact both are equally derivative of a fundamental value assumption. Toxins are kinds of harm, and although it may be factual that a particular animal or plant is harmed by plant X, plant X may also benefit other animals or plants, as well as the ecosystem.

BT CORN CONTINUED: RESISTANCE CREATED BY GM PLANTS?

Naturalists Hubbell and Welsh think that continued planting of Bt corn and Bt cotton will create weeds resistant to Bt:

> In fact, by focusing weed management efforts on a single strategy, herbicide resistant crops can increase weed problems by speeding up the development of herbicide resistant weeds. Monsanto has argued

> (1997) that there have been no verified cases of weed resistance to glyphosate [Roundup's ingredient]. However, Sindel (1996) has documented glyphosate resistance in annual ryegrass in Australia. Further, Gressel (1996) provides theoretical evidence that glyphosate resistance in weeds is not unlikely and that "there are few constraints to weeds evolving resistance to glyphosate." Farmers planting herbicide tolerant crops have little incentive to use multiple weed control strategies. Instead, they have incentives to intensively apply a single chemical, leading to natural selection of weeds tolerant of that particular herbicide. In fact, farmers planting Monsanto's Roundup Ready soybeans are required to sign a contract proscribing them from using any herbicide other than Roundup.[22]

Second, they also argue that herbicide-tolerant crops may themselves become weeds. If true, then farmers would have difficulty controlling such weeds with traditional chemical means. Furthermore, by giving a traditional crop a competitive advantage over other plants in an ecosystem, Hubell and Welsh claim that agribusiness could in effect create "exotic" invasive species similar to purple loosestrife and kudzu.

Hubbell and Welsh think Boll Guard cotton will create similar problems. Over the last decades, budworm and bollworm, two chief pests of cotton, have become resistant to synthetic pyrethroid insecticides.[23] Despite repeated sprayings of such pyrethroids, these pests inflicted major damage on cotton crops and the pesticides had significant effects on waterways and surrounding insects. Boll Guard cotton, with built-in Bt, enabled farmers to dramatically reduce spraying of pyrethroids and opened the door to nature-friendly, integrated pest management (IPM).

However, Hubbell and Welsh claim that if Bt is merely swapped out for pyrethroids, without adoption of IPM strategies, then budworm and bollworm will soon build up resistance to the Bt in Bt cotton. Rissler and Mellon of the Union of Concerned Scientists claim that such resistance has already occurred.[24] If so, then traditional insecticides will have to be brought back, and Bt will have to be sprayed externally, too. All in all, Hubbell and Welsh think that the strategy of swapping Bt cotton for pyrethroid spraying "is just a delaying tactic" against the same bad outcome as before: "resistant pest populations and ever-increasing chemical applications to realize decreasing levels of damage control."[25]

If these naturalists are correct, then Bt corn, cotton, and soybeans will eventually bankrupt farmers, who will have to pay for more pesticides applied more frequently; create a nightmare for the environment with ever more tons of pesticides going into waterways and surrounding ecosystems; and breed superpests resistant to genetically modified and traditional pesticides. As Mellon and Rissler ominously second Hubbel and Welsh:

> The problem with transgenic Bt plants is that they will increase the exposure of pests to Bt, especially where Bt is continuously expressed in a plant throughout the growing season. This is significant because long-term exposure to Bt toxins promotes development of resistance in insect populations. This kind of exposure could lead to selection for resistance in all life stages of the insect pest on all parts of the plant for the entire season. (McGaughey and Whalon, 1992)[26]

I think Hubbell and Welsh, as well as Mellon and Rissler, make a good argument against GM crops, but not a knockdown one. The environment, be it pests or plants, will adapt to plants with new genes. Rather than "resistance is futile," here "resistance is inevitable." These four critics correctly predict that some pest populations eventually will develop resistance.

But what else is new? How is that different from what happens with use of chemicals? And if we don't use either chemicals or GM techniques, and since we can't grow enough food for ourselves or the world organically, what alternative do we have?

On the other hand, if GM plants are just tools—as I've argued in this book—then we can develop better and better tools. After all, the first saw invented wasn't anything like the best saw used today.

Indeed, insertion of specific genes gives us new and more precise tools to counter resistance: as pests adapt, so will our tools.

Moreover, the new emphasis on IPM can allow us to mitigate the problems of resistance, maximize the effectiveness of GM plants, and even mix in occasional use of organic techniques. Moreover, all this can be done without excessive federal or state regulation because it's the farmer's income and capital (his land) that are at stake with failure. No sane farmer wants to encourage pest resistance.

TERMINATOR GENES CAN HELP THE ENVIRONMENT

One of the major objections that environmentalist groups had against genetically modified crops is transgenic contamination. Marc Lappé and Britt Bailey think that the evil corporations that formerly gave us DDT, Agent Orange, and PCB's have retooled into biotechnology and lament,

> Few of the seeds created by genetic engineering are non-propagative, and hence nonviable by design. Fertile transgenic pollen can and has escaped to contaminate weedy species. Even this unfortunate side effect could have been offset by encoding the genes for male sterility or selfing genes that prevent fertilization. Without such built-in protections, gene flow between transgenic and native species remains a disturbing possibility.[27]

So what did these same naturalists and egalitarians do when the evil corporations developed plants exactly as they wanted to prevent transgene transfer? They almost rioted in condemning the companies for this move and explained that the development could come from the vilest of motives.

Naturalists now proclaimed that these autosterilized seeds would lead to the death of all plant life when the sterilizing transgenes escaped into the environment via natural pollination. Moreover, farmers would be forced into perpetual servitude because they would have to buy seeds each year from these companies.

Progressives hold a real ace here because terminator genes are the perfect solution to the problem of interplant gene transfer. Preventing the reproduction of contaminated plants (and thus the continuation of the genes), makes it nearly impossible for the transgenes to escape. Although DPL scientist William Hugie admits that "yes, it could happen," he also believes that its negative effect will be so small as to go unnoticed, especially with its positive effects in controlling genetic pollution.[28]

IS BIO-CONTROL BETTER? NOT NECESSARILY

One would think that a good way to control a Russian wheat aphid infestation would be to bring in a batch of, say, ladybugs. But as we saw with killer bees, traditional natural methods are not always good if not used carefully and if not integrated wisely with other tools.

For example, ladybugs don't hurt the environment and they naturally prey on aphids anyway. But what happens with the ladybugs when the aphids are gone? Naturally, they don't just prey on one species of aphid, but on other organisms as well. The result can be a dominant species let loose in huge numbers in an environment with perhaps no natural population control. With an abundance of food, the exotic species multiplies rapidly and will require more food, which will most likely be native, regional species, some of which may be endangered. The larger population learns to feed off other indigenous species and, with no natural controls, takes over the ecosystem. The old pests are gone, but a new pest claims the niche.

Consider a different scenario. To eradicate a disease spread by a blood-sucking insect, rather than introduce a new potential pest, a scientist genetically alters the bug so that it is no longer physiologically able to carry the microorganism that causes disease. Inundating a "wild" population with these altered insects gradually genetically wipes out the ability to carry the pathogen, resulting in a pesky, but not deadly, population of insects. From an agricultural point of view, pest populations can also be inundated with sterilized populations to radically decrease the efficacy of mating. Gradually, the only bugs left are sterile ones, which cannot reproduce and simply die.

From a naturalist viewpoint, the insertion of a toxin-producing gene into viruses or bacteria gives the toxin a new path for transgenic movement. Also, the populations of viruses and bacteria, once released, are nearly impossible to control, and it is just as impossible to prevent mutations and genetic spread of the toxin.

BT CORN, AGAIN

In response to John Losey's study on Bt corn and monarch butterflies, Progressives retorted that the study did not represent the actual conditions under which Bt corn is grown. First, they said, it is not surprising that Bt acts against larvae of butterflies, since they are members of the same lepidoptera order as the European corn borer. Second, monarch butterflies lay larvae only on milkweed. Growers do everything they can to rid their fields of all weeds, including milkweed, so there are not going to be many monarch larvae in the middle of cornfields.

Third, corn has heavy pollen, with 90 percent falling within six yards of the outer edge of the field, meaning it does not get blown far by wind or taken far by insects. Losey said in his original article that milkweed grows well in "disturbed" areas such as roads along cornfields. So some monarch larvae on roads along and through cornfields would be affected. A previous, largely overlooked study of Bt corn pollen, milkweed, and affects on monarchs *in the field* (Losey's was in a laboratory) showed only a 16 percent death rate of monarch larvae in milkweed pots left inside cornfields (Losey got a 44 percent death rate inside).[29]

Fourth, progressives claimed that in order for the most sensitive monarch caterpillars to start to die, Bt pollen would have to be ingested in amounts ten to fifty times the amount that would be ingested on the most heavily pollinated field.[30] Progressives think that the amounts Losey used to kill monarch larvae far exceeded the amounts real larvae would feed on.[31] Fifth, monarchs do not like milkweed covered with Bt pollen and seek out nonpollinated milkweed. Indeed, monarch larvae dislike milkweed leaves covered with *any* pollen, be it Bt or non-Bt.[32]

Sixth, Bt pollen does not stay for long on leaves of milkweed; corn only sheds in a particular field for around a week. From this, combined with the fact that corn pollen is heavy and the fact that farmers don't want milkweed around their cornfields, one may conclude that few monarch larvae will actually feed on Bt pollen in real fields. Moreover, the few larvae that are exposed will probably not be exposed in high enough amounts to do damage.

In their original study of Bt corn pollen on monarch butterfly larvae, the EPA had come to the same conclusion in allowing Bt corn seed to be sold. The EPA thought that only a small amount of milkweed would be covered by Bt pollen and only a small percentage of these would have monarch larvae at exactly the right time to cause deaths.

Perhaps the most telling point of all, however, was the lack of something in Losey's study, the omission of which reinforces the elementary lesson of all scientific studies. This is the first thing every student learns in any science class and its omission in Losey's study, when it could have been easily included, speaks volumes to the motives of the researcher. Why were there no controls? Why was there no corn sprayed, for comparative purposes, with the normal chemicals used to control insects and weeds? How do we know that such chemicals do not also hurt monarch larvae at particular points in time?

That no control was used says to me that Losey was not interested in finding out whether growing Bt corn hurt monarchs worse than the traditional practice of using chemicals: he simply wanted to show damage. This got a lot of press, but it isn't good reasoning or good science.

REDUCTION OF USE OF CHEMICAL PESTICIDES

Progressives also argue that Bt corn has benefits that compensate for a small number of deaths of monarch larvae. Because genes for

Bacillus thuringiensis reside inside the corn, smaller amounts of chemical spray need to be applied to whole fields to prevent damage from European corn borers. For progressives, this seems very beneficial, in part because chemical pesticides are expensive to farmers, who want to save costs and maximize profits.[33] Also, such a reduction in pesticides prevents the deaths of other insects and organisms that the environment needs and that would otherwise be killed by chemical, synthetic pesticides.

Naturalists remain unconvinced. They claim that *more*, not less, chemical pesticides are used on Bt corn. In a recent study, scientists concluded that no firm evidence exists that Bt soybeans and cotton actually reduce the need for pesticides on crops.[34] The evidence was inconclusive. Even so, there was also no evidence that Bt corn requires more pesticides.

Naturalists retort that the European corn borer will develop resistance to Bt corn. Especially if corn growers repeatedly plant Bt corn in large monocultures, insects will develop resistance to Bt much faster.

This objection by naturalists (e.g., the Environmental Defense Fund) has been taken seriously by North American farmers and the USDA, so to prevent resistance the American and Canadian departments of agriculture required farmers growing Bt corn to plant at least 20 percent of their corn with conventional strains, and if Bt grown was planted near similarly enhanced cotton (Boll Guard cotton), only 50 percent of the field could be Bt corn.[35] Notice that critics cannot have it both ways here: they cannot claim that the system is unresponsive and won't protect them when it does respond.

Naturalists also retorted that one strain of Bt corn pollen—Novartis Max 454—expresses forty times more Bt protein and hence has much higher mortality rates to monarch larvae.[36] If true, Progressives can agree that not every farmer should be free to plant every variety of Bt corn.

END COMMENT

Of all the critiques of GM plants, worries about environmental damage carry the most weight. But as we saw in the previous chapter, preserving the environment (as it's been culturally constructed at this point in history) is neither an absolute value nor the only value. Interests of humans matter the most, especially starving humans.

Any introduction of change in the environment has the potential for disaster. That is the root worry behind the naturalist critique of GM crops. What the naturalist ignores, in focusing on GM plants, is that modern commerce, human vacationers, and international flights constantly move plants and animals around the world (despite the best efforts of inspectors at ports and airports to prevent it).

As we saw with killer bees, most such introductions, even when intentional, were crude hit-or-miss genetic affairs. Modern GM techniques offer a more precise, safer way of doing the same thing: not "killer tomatoes," but maybe tomatoes with smiley faces on their skins.

There will be mishaps. That is inevitable. There will be development of resistance and unforeseen consequences. At the first sign of the first problem, alarmists will say the heavens will open and boll weevils will descend as punishment for tampering with Mother Nature. What else is new?

In the end, I believe that GM plants will be just better tools for the many ways that humans use plants: for pretty gardens, for profitable farming, and for the aesthetic value of wildness around them. How these tools are used is up to us. By themselves, they are neither good nor evil, just tools.

9

Six Concluding Reflections

Whoever could make two ears of corn, or two blades of grass grow upon a spot of ground where only one grew before; would deserve better of Mankind, and do more essential service for his country, than the whole race of politicians put together.

—*The King of Brobdingnag,* Gulliver's Travels, *1727*

1: "PIG CITY" AND GM PIGS

Our global, industrial food system is starting to become absurd, as shown in some compromises made between the primitivism of naturalists and the laissez-faire of globalists. One example is a proposed "Pig City" in Holland.

The Netherlands is famous for exporting tasty hams, hence a proposal there to build a massive high-rise (maybe eighteen stories) office-style building designed to house up to 300,000 pigs. The facility would be near an urban port, making it easy to export hams, bacon, and sausage, or just the whole pig.

Because land is scarce in Holland, and because hog farms both take up a lot of space and create a lot of environmental waste, this proposal was almost realized. (Raised naturally, a pig needs a hundred acres or more to roam and root.)

But an eighteen-story Pig City strikes me as a reductio ad absurdum of our food system. Fortunately, the outbreak of hoof-and-mouth disease in Europe impressed on Dutch officials the obvious dangers of such a facility in the face of such an outbreak.

Even if it were hidden from public view (as I'm sure it would be, like most hog farms), this seems too much to me like one of the those vertical prisons in downtown American cities. Also, if a vertical hog farm with 300,000 pigs on the edge of a city is repulsive, why isn't one spread out horizontally but hidden in some rural part of the South? It seems to me both are.

The more one learns about such things, the more one loses one's taste for pork. I believe people ought not eat pig flesh on moral grounds to try to reduce the suffering of these intelligent animals. But it's easy just to learn some facts about pigs and hence (at least for me) to lose the desire to eat them.

Would seeing thousands of acres of pineapples grown in Hawaii have the same effect on me? I don't think so. Monocultures of plants, however they may or may not threaten biodiversity and be risky business practices, don't cause sentient beings to suffer.

But should we allow creation of new animals that might be more efficient to raise as food? Creation of GM plants raises few issues, but what about GM cows, pigs, and chickens? As mentioned, plants are not sentient and hence changing some of their genes, or mixing their genes in crossbreeds, poses no moral problems by creating suffering in plants.

Animals are different. For example, pigs are known to be very intelligent, much more so than cats or dogs. Wouldn't it be more humane to create a very retarded pig who was content to sit around all day eating, i.e., who didn't suffer as much from the confinement of huge hog farms or a "Pig City"?

I am of two minds. It would be better not to eat pigs and products derived from them. On the other hand, a good utilitarian interested in animals might accept the fact that consumption of beef, pork, and chicken has actually *risen* in North America over the last decades, despite more people becoming vegetarians (of course, there are also more people in North America now, but this alone does not explain the rise).

Much as I hate to admit it aesthetically, I can see no moral reason for creating barely sentient pigs destined to become human food, especially if strong evidence exists that such semi-pigs had strongly diminished ability to suffer. Such GM pigs would be more like domesticated chickens and cows, which have been bred over many generations to be relatively mindless.

I realize that this conclusion will infuriate purists who say we should just press for abolition of eating all kinds of animal flesh. But if practical ethics is to confront the real world, practical suggestions must be made. This seems to me one that could reduce the suffering of the millions of animals raised to be consumed for human pleasure.

2: LABELING GM FOODS?

Labeling GM food seems to be a reasonable compromise. Even if scientists insist on its safety, shouldn't such food be labeled to give information to consumers? Isn't labeling like informed consent in medicine? Even if we can't agree on standards of absolute safety, i.e., if we can't agree at the level of content, can't we agree that giving out more information is better than giving out less, i.e., can we agree on a process?

When I first began to research GM food, I thought labeling them made sense. Now I don't. In context, enforcing such proposals is unduly alarmist and sends the message that there is some special danger here that consumers must be warned about. In contrast, the nonlabeling of other foods, where the plants derive from crossbreeds and new kinds of imported foods, sends the wrong message that these foods are perfectly safe.

The recent move to require a U.S. government standard of labeling for organic food is part of this issue. Consumers are likely to interpret food labeled as "organic" as safer and more nutritious than food not so labeled. If some food is labeled as containing "genetically modified ingredients," then the same consumers may scorn such food as dangerous and nonnutritious.

Lurking beneath the labeling issue is the much more potent issue of full disclosure about all products that pose any risks to consumers. If we're going to label, we shouldn't pick on plants. What about gelatin, used in Jello and in many capsules that liquid vitamins come in (such as vitamin E)? Should it be disclosed that gelatin can come from "animal products" that may contain spinal remnants and, as such and in some countries, subject consumers to mad cow disease and other TSEs?

I believe that calls for universal labeling of GM foods are like calls in the early days of the AIDS crisis that we test all Americans, or all applicants to medical and nursing schools, for HIV. Ignorance was behind such proposals, and the correct response to ignorance is not to validate it by making laws based on it, but to counter it with facts and knowledge.

3: WENDELL BERRY'S AMERICA

American naturalist Wendell Berry writes eloquently, passionately about returning America to a land of small, self-sufficient, family-run farms where people treasure the land, embrace manual labor, take pride in producing healthy food, and live debt free. He thinks such a farm culture sustains moral virtues such as thrift, hard work, and pride in one's land and its produce.

Berry hates agribusiness, specialization in farming or work, cities, and suburbia. He also doesn't like shopping malls, Starbucks coffee (or any other chains), interstates, jobs in office buildings, or people who aren't really "from" any state or region.

While it is nice that some farmers loved their work, so much that they "lived in their barns,"[1] few ever did and even fewer would today. Speaking as one who once slept for one night in a barn, I think romanticists such as Berry always leave out the ticks, fleas, chiggers, and rats that inhabit barns and their surroundings. His whole picture of farm life is so romanticized that it might be from one of those newly popular coffee-table books with wonderful, color photos of old farms and barns.

"A healthy culture is a communal order of memory, insight, value, work, conviviality, reverence, aspiration," he writes. "It reveals the human necessities and the human limits. It clarifies our inescapable bonds to the earth and to each other. . . . [A] healthy farm culture can be based only on familiarity and can grow only among a people soundly established upon the land; it nourishes and safeguards a human intelligence of the earth that no amount of technology can satisfactorily replace. The growth of such a culture was once a strong possibility in the farm communities of this country. We now have only the sad remnants of those communities."[2]

Much of this is true, especially the part about passing down knowledge of how to work with the land from generation to generation, but much of it is also sad, much sadder than Berry indicates. For what he never mentions is that farm life was often too hard, often very boring, and that almost all those millions who left farming (like my own parents and grandparents) left it *voluntarily.* Now only the rich psychiatrist, CEO, or professional athlete can wax eloquent about staying in touch with his farming roots and with "the Land." For most young people, the possibility of living a well-paid, satisfying life in the country has long since disappeared: even for those who try to keep family farms going, divorce is high and many single men who've inherited a farm find it difficult to find a woman willing to marry into such a life. Moreover, almost all those who spout ideas like Berry's or believe that the world of Garrison Keillor's *Prairie Home Companion* still exists are hypocrites: while those who champion them cite the self-reliance and self-sufficiency of small farmers, almost all farms today depend on welfare checks from the federal government to pay bills, and without such checks would go bankrupt.

4: "FOOD IS A WEAPON"

"Earl Butz," the former U.S. secretary of agriculture, once said that "food is a weapon," by which he meant that a nation can't just charitably transfer it from its farmers to those who are starving.[3] Instead, we must transfer food to those whose policies align with ours. That statement chafes a lot of people, who think food should be used compassionately, not in politics and wars. We should simply *give* food, or what people need to grow their own food, to others.

That policy doesn't work for two reasons. Altruism is limited, especially that available to masses of strangers on the other side of the globe. Even where food is transferred, someone pays farmers for the food transferred—be it the U.S. government or the World Council of Churches or Oxfam.

Altruism is easier when someone else pays. One of the main reasons why the U.S. government aided starving people in the past is that such food transfers benefited agribusiness in the 1980s and 1990s by creating new markets. Powerful congressmen could help agribusinesses contributing to their reelection campaigns and claim the moral high ground while doing so.

In real food policy to end starvation, one must respect the reality behind the remark that "food is power." Not everyone in the world is nice. Dictators control people against guerilla insurgents by controlling access to food. Professor Sen is correct: food is power, and it is used as such in many countries where people starve.

And in such countries, neither charitable transfers of food and supplies nor miraculous revolutions in food production (GM foods or the previous Green Revolution) will alone end starvation. GM food is only a tool, and it is one that could be used by dictators in evil ways, e. g., finding a lethal virus that would destroy the monoculture of one's enemy. But just like tools for digging wells, harvesting grain, or tilling the soil, GM foods should not be banned for being a new tool and possibly susceptible to such uses. Let's just see it for what it is and hope that people use it benevolently.

5: VANDANA SHIVA'S INDIA

Although trained as a physicist, Vandana Shiva writes as an agricultural, Hindu theologian who wants to preserve each species as a natural kind and maintain an ancient India where 80 million cows formed the background both of the religion and a farming system where women did most of the milking, feeding, and recycling of manure. Perhaps the world's most famous critic of GM food and industrial food systems, Shiva has never acknowledged any real alternative to returning to the past. Nor has she acknowledged over the past decade the importance of Borlaug's Green Revolution in saving India from starvation.

I want to mention something here that Vandana Shiva and Wendell Berry have in common and to offer one explanation: they both write very well and you always know where they stand; indeed, you can predict where they stand on agricultural issues. Ms. Shiva and Mr. Berry are undoubtedly very smart and very knowledgeable about agricultural values, but it is much easier than normal to write well when defending a nostalgic past and criticizing colossal monsters such as "international agribusiness."

Thinking simplistically allows one to write with passion, without qualification, and with a style that suggests one is battling Absolute Evil. As readers of Jeremy Rifkin and Leon Kass also know, it is easier than usual to be eloquent when defending the status quo against technological change, for fear of such change taps into the Accident Story, our most ancient myth about the evils that befall humans who reach too far.

But such thinking and writing distort the complicated truths about humans today. For example, I don't know any Indian girls who want to be milkmaids, but I know a lot who want to be physicians. In India's high-tech Bangalore, where programmers compete to build better systems and occasionally make world-class breakthroughs (such as creation of the web-based e-mail program Hotmail), no one wants to return to a life of long hours of recycling cow manure. Vandana Shiva does not speak for the real India.

6: FOOD AND PHILOSOPHY

Whether it's moral or immoral to use food as power, in the parts of the world where it's used this way, food is a life-or-death issue. So whether we *use* food as power is one thing, but we definitely should *think* more of food in terms of power.

And food has power in many different ways. When food is considered as basic to life, in countries where some people are starving or malnourished, who controls food controls life and death, health and disease. As a commodity in world trade, food is economic power, and just because North America grows food easily, it should not ignore how such power is used or abused. As Jared Diamond explains in his Pulitzer Prize–winning *Guns, Germ, and Steel*, growing food easily and abundantly means being very lucky about lots of factors coming together.[4] It is easy to delude ourselves with the racist belief that the technology or industriousness of North Americans and Europeans are responsible for abundant food on those continents compared to central Africa, but the truth is much more complex.

Food is also cultural power. The symbolic significance of GM food seems to be why Europe has drawn a line in the sand and against it. I think this is much ado about nothing, because GM food creates many more culinary possibilities than it eliminates (organic is limiting: think what could be done if all fresh produce could be kept fresh and taste fresh, really taste fresh, for an extra few days or even a week). Maybe some elitists think, much as I did as a juvenile when I made my hike in the Wind River Mountains, that only people who work hard for fresh produce (or are willing to pay extra) deserve to taste such food. But these are old, untested, elitist assumptions coming to the surface that really won't bear public defense.

A lot depends on which direction food policy points. Our new Rousseaus have no real plan for the suffering peoples of countries of the developing world. Critics only see the by-products of change: old crafts abandoned for jobs in Nike factories, old forms of entertainment lost when MTV and night clubs entice youth away. Sad as these changes may be, human history has shown only one model for raising the living standard of peoples in developing countries. As Ronald Bailey writes admirably in *Reason* magazine,

> As much as the neo-Luddites might wish it otherwise, there simply is no other social and economic model of lifting hundreds of millions of people out of poverty than what might be called democratic, technological capitalism. If one wants effective sanitation, improved medicine, a steady food supply, convenient transport, and cheap and easy communications, there is no alternative to technologically robust, market-based societies. To the arguable extent that countries worldwide are becoming more similar, it is not because corporations are imposing some uniform set of goods and services, but because human beings share a similar set of needs and wants.[5]

New York University has just started a new Ph.D. program in food studies, so interest must be growing in an interdisciplinary approach to thinking about food. Political scientists, anthropologists, sociologists, historians, philosophers, bioethicists, and economists should all be writing about food as a social, political, and ethical subject. As I hope this book has shown, the topic is more important than one for just menus and recipes.

Appendix

Groups Advocating Food Policy

Note: When I began my research on GM food, I was confronted by a bewildering number of centers, institutes, and organizations with acronyms. It took me a long time to figure out who was whom, and what each group represented. I still don't understand groups, how they are funded, or their motives. It is also difficult at times to distinguish an organization of people from a web site run by an individual. The following is not an exhaustive list but a place to start in understanding these groups.

American Corn Growers Association (ACGA), not the main professional association of corn growers (which is the National Corn Growers Association), but an anti–GM food group with ties to Jeremy Rifkin and Friends of the Earth.

American Society for Cell Biology, the professional society of this standard department in medical schools, which supports research on genetically modified foods and crops.

AstraZeneca PLC, Anglo-Swedish agribusiness and fourth-largest supplier of Golden Rice. The company that in 2000 gave Golden Rice free to poor farmers to prevent vitamin A blindness.

Center for Food and Agricultural Research (CFFAR), pro–GM food think tank.

Center for Global Food Issues, part of the Hudson Institute. Very pro–GM food. The director is Dennis T. Avery.

Center for Food Safety (CETOS), located in a suite on Pennsylvania Avenue in Washington D.C., easily confused with the FDA's Center for Food Safety and Nutrition. CETOS is a naturalist-egalitarian advocacy group whose board members include Sheldon Krimsky, a Tufts University professor; Tony Kleese of the Sustainable Farming Program; Melanie Adcock, the program officer for the Foundation for Deep Ecology; Margaret Mellon, director of the Union of Concerned Scientists' project on Biotechnology and Agriculture; chef Alice Waters; Michael Sligh, director, Rural Advancement Foundation International (see below): and Abby Rockefeller.

Community Supported Agriculture (CSA), a program of the USDA. See www.nalusda.gov/afsic/csa/.

Competitive Enterprise Institute, based in Washington, D.C., has a food safety policy director, Gregory Conko. Very pro–GM food.

Council for Agricultural Science and Technology (CAST), an industry-based organization dispensing "science-based" information on food and agricultural policy, not to be confused with another CAST, the Center for Applied Special Technology, which helps people with disabilities.

Very important note: CAST has a web page with a listing of hundreds of acronyms in food science and agricultural policy at www.cast-science.org/acronyms/acronyms.htm#I.

Council for Biotechnology Information, sponsored by six big biotech companies, including Aventis, Dow Chemical, Monsanto, Novartis, Zeneca, and DuPont.

CropChoice, a coalition of anti–GM food groups that include the Center for Food Safety, Food First, Demeter Association, Family Farm Defenders, Greenpeace USA, National Family Farm Coalition, Maine Organic Farmers and Gardeners Association, the Sierra Club, New York Coalition for Alternatives to Pesticides, Institute for Agriculture and Trade Policy, and Sustain: The Environmental Information Group.

Earth Liberation Front, radical environmental group.

Economic Strategy Institute, proglobalization and pro–GM foods, based in Washington, D.C.

Environmental Defenders Office, Victoria, Australia, antibiotechnology in Australia.

Environmental Defense Fund, biologist Rebecca Goldberg often speaks for this organization, which is anti–GM crops.

Environmental Media Services (EMS), an activist, web-based source for journalists.

Food and Drug Administration (FDA) of the United States.

Food First, radical food organization with links to Vandana Shiva and Mark Lappe, also called Institute for Food and Development Policy.

Food Industry Environmental Network (FIEN), pro-industry information service on GM foods, biotechnology, and environment.

Friends of the Earth (FoE), a radical environmental group based in England that is very anti–GM food. Also has groups in other countries, under *Friends of the Earth International (FoEI).*

Greenovation, spin-off of the University of Freiburg in Germany founded in late 1999 as a for-profit company doing research for private companies in agriculture and pharmacology.

Grocery Manufacturers of America (GMA), Brian Sanson, spokesperson.

Hudson Institute, founded by Herman Kahn, Max Singer, and Oscar Ruebhausen in 1961 in New York City. In 1984, after Kahn's death, moved to Indianapolis, Indiana. Thinks of itself as promoting unconventional thinking about the future. Protechnology and antiregulation. See www.hudson.org/abouthudson.cfm.

InterAcademy Council (IAC), based in Amsterdam and founded in 2000, hopes to bring scientists, engineers, and medical experts together to advise the United Nations and the World Bank.

Interfaith Center on Corporate Responsibility, a coalition of three hundred "socially responsible" organizational investors. Anti–GM crops and anti–GM food. Includes Sisters of Notre Dame de Namur in Belmost, California, and its leader, Susan Vickers.

International Food Information Council (IFIC), industry-supported organization providing information to leaders of food and agricultural industries. Web site: ificinfo.health.org.

International Potato Center (CIP), a nonprofit agency working out of Lima, Peru, to develop better potatoes without natural toxics found in potato tubers from glycoalkaloids.

Institute for Agriculture and Trade Policy, (IATP), campaigns against patents on genes and for labeling of GM food.

Institute for Food and Development Policy, Peter Rosett, director, author of *World Hunger: Twelve Myths* (egalitarian on famine). See Food First.

Institute of Food Science & Technology, in United Kingdom. Ralph Blanchfield edits their web-based newsletter.

National Center for Food and Agricultural Policy (NCFAP), in Washington, D.C., issued a report in 2000 on why growers choose GM soybeans. Provides data of use to discussions of farm policy.

Physicians and Scientists for Responsible Application of Science and Technology (PSRAST), anti–GM food group that cites questionable research against the safety of GM food.

Research Foundation for Science, Technology, and Ecology (RFSTE), in India. Anti–GM food and crops. Accused the United States of dumping GM food on starving people.

Rockefeller Foundation, an old, famous philanthropic organization funded originally by John D. Rockefeller's oil money (Esso, later Exxon) and mandated to help the excluded peoples of the world have better lives. Has focused in past decades on birth control and improved crops to prevent starvation.

Rural Advancement Foundation International (RAFI), Michael Sligh, director, board member of CETOS. Spearheaded opposition to terminator seeds in GM crops. Hope Shand of RAFI is credited with the invention of the term "terminator genes," an effective PR tool. Also uses the term "traitor technology" to describe what Monsanto called "GURTs" or genetic use restriction technology.

Sustain: The Environmental Information Group, aims to mobilize the public against biotechnology, especially by organizing rallies outside hearings at public agencies.

Syngenta AG, new company formed in 2000, when AstraZeneca and Novartis merged their agrochemical and seed businesses.

Tides Foundation, nonprofit foundation seeking "to make the world a better place" by opposing GM food, the death penalty, and nonliving wages for workers. See www.tides.org/foundation/index.cfm.

United States Department of Agriculture (USDA).

Zeneca Agrochemicals, crop and plant division of AstraZeneca.

Notes

CHAPTER 1

1. www.cdc.gov/ncidod/eid/vol1no2/feng.htm. For documentation from the beef industry, especially about the greater dangers of *E. coli* O157:H7 from fruits and vegetables, see www.beef.org/library/factsheets/fs_e_coli.htm.

2. Peter Feng, "Escherichia coli Serotype O157:H7: Novel Vehicles of Infection and Emergence of Phenotypic Variants," Centers for Disease Control, www.cdc.gov/ncidod/eid/vol1no2/feng.htm.

3. Barry Yeoman, "Dangerous Food," *Redbook,* August 2000, 123.

4. www.thisislondon.co.uk/dynamic/news/business/top_direct.

5. Christina Pope, "Fact Sheet: E. coli O157:H7," Beef Industry Food Safety Council (BIFSCo), www.beef.org/library/factsheets/fs_e_coli.htm.

6. "Warning: Organic and Natural Foods May Be Hazardous to Your Health," Knight Ridder/Tribune News Service, 2 May 2000 (Internet edition).

7. Sandra Blakeslee, "Pesticide Found to Produce Parkinson's Symptoms in Rats," *New York Times*, 5 November 2000 (Internet edition).

8. From a posting by medical toxicologist Alan H. Hall on the AgBio Listserv, dated 9 May 2000.

9. "Seeds of Change," *Consumers Reports*, September 1999, 43.

10. Michael Pollan, "How Organic Became a Marketing Niche and Multi-billion-Dollar Industry," *New York Times Magazine*, 13 May 2001, 32.

11. Marian Burros, "Mainstream Organics: Britain Stocks Up," *New York Times*, 26 June 2000, B11 (national ed.).

12. Norman Borlaug, "Taking the Gm Food Debate to Africa: Have We Gone Mad?" open letter to the editor, *Independent Newspaper*, London, 10 April 2000.

13. Jim White, "Organic Foods More Healthful," *Albuquerque Journal*, 2 August 2000, C1.

14. James Meek, "Prince Courts Controversy As He Places the Nature of God above the God of Science," *Guardian*, 17 May 2000, Internet edition.

15. James Meek, "Duke Challenges Sceptics over GM Food," *Guardian*, 7 June 2000 (Internet edition).

CHAPTER 2

1. Mark Lappe and Britt Bailey, *Against the Grain: Biotechnology and Corporate Takeover of Your Food* (Monroe, Me.: Common Courage, 1998), 12.

2. Lappe and Bailey, *Against the Grain*, 9

3. Donald G. McNeil, "European Conference on Food," *New York Times*, 29 February 2000.

4. McNeil, "European Conference on Food."

5. McNeil, "European Conference on Food."

6. McNeil, "European Conference on Food."

7. Timothy Egan, "'Perfect' Apple Pushed Growers into Debt," *New York Times*, 4 November 2000, A1, 9.

8. Europeans feel they ought to patronize European movies when out on a weekend night, but in fact they usually opt for the American

movies. Even so, such movies do not usually portray America favorably: the top movies in April 1999 in Europe were *The Green Mile, American Beauty,* and *Erin Brockavitch*—and think of the underlying messages conveyed in those films about American life!

9. Suzanne Daley, "More and More Europeans Find Fault with US: Wide Range of Events Viewed As Menacing," *New York Times*, 9 April 2000, A1.

10. Daley, "More and More Europeans Find Fault with US," A1.

11. Michael Fumento, "Crop Busters," *Reason*, January 2000.

12. David Sapsted, "Ex-minister Denies Damage and Theft of GM Trial Crop," *Daily Telegraph*, 4 April 2000. Internet edition.

13. David Brown, "Farmers Accidentally Sow 30,000 Acres of GM Crops," *Daily Telegraph*, 24 May 2000.

14. AP, "Greenpeace Intercepts Ship Carrying GM Crop," 25 February 2000.

15. Ray Losley, *Chicago Tribune*, carried in *Birmingham News*, 25 May 2000, 4A.

16. Losley, 4A.

17. Burros, "Mainstream Organics," B11.

18. Suzanne Daley, "French Rally around Unlikely National Hero," *New York Times*, 2 July 2000, A1; Pierre-Antoine Socuhard, "Anti-global Ire Up at 'McTrial,'" Associated Press, *Birmingham News*, 2 July 2000, A16.

19. *Seattle Times*, 6 December 1998, front-page story summarizing events at the WTO meeting.

20. Sam Howe Verhovek, "Trade Talks Start in Seattle despite a Few Disruptions," *New York Times*, 30 November 1999, A14.

21. Kenneth Klee, "The Siege of Seattle," *Newsweek*, 13 December 1999, 30–39.

22. *Wall Street Journal*, 11 November 99, A27.

23. Francis Fukuyama, *Wall Street Journal*, 17 December 1999.

24. "The WTO: The Villain in a Drama It Wrote," *Wall Street Journal*, 6 December 1999, A1.

25. "The WTO," *Wall Street Journal*, A1.

26. "Sugar Story," *New Orleans Times-Picayune*, 6 December 1999.

27. Elizabeth Becker, "Far from Dead, Subsidies Fuel Big Farms," *New York Times*, 14 May 2001, A1.

28. Rachel Smolkin, "Frankenfoods: Engineered Foods Hot Topic at Talks," Lead story, *Birmingham Post-Herald*, 3 December 1999, A1.

29. "The WTO," *Wall Street Journal*, A1.

30. "The WTO," *Wall Street Journal*, A1.

31. Andrew Pollack, "Talks on Biotech Food Today in Montreal Will See U.S. Isolated," *New York Times*, 24 January 2000, A10 (national ed.).

32. Pollack, "Talks on Biotech Food," A4.

33. John Burgess, "U. S. Accepts Deal Guiding Genetically Engineered Products," *Washington Post*, 30 January 2000, 1A; reprinted in *Birmingham News*, same date, 9a.

34. Fumento, "Crop Busters."

35. Sam Howe Verhovek with Carol Kaesuk Yoon, "Fires Believed Set As Protest against Genetic Engineering," *New York Times*, 23 May 2001, A1.

CHAPTER 3

1. Wendell Berry, *The Unsettling of America: Culture and Agriculture* (San Francisco: Sierra Club Books, 1986), vi.

2. Mae-Wan Ho, "GM Foods and the Luxury of Choice," *Times* (London), 21 March 2000.

3. C. S. Prakash, "Nobel Prize Winners Endorse Agricultural Biotechnology," 7 February 2000. Press release, www.agbioworld.org/pr/watson.

4. Quoted in Thomas DeGregori, "Genetically Modified Nonsense," Executive Summary, www.wepa.org/html/near218a.htm.

5. "Why the West Must Swallow Gene Foods," *The Observer*, 23 January 2000.

6. Jeremy Rifkin, *BioTech Century: Harnessing the Gene and Remaking the World* (New York: Tarcher/Putnam, 1998).

7. Mae-Wan Ho, "EU Green Scientists Slam Biotech Foods" (presented at the Green Conference on Genetic Engineering, Brussels, March 9, 1998). See www.purefood.org/ge/geslam.html.

8. John Rawls, *A Theory of Justice* (Cambridge, Mass.: Harvard University Press, 1971).

9. Vandana Shiva, *Biopiracy: The Plunder of Nature and Knowledge* (Boston: South End, 1997), 119.

10. Satish Kumar, interviewed in "Soul Man," *New Scientist*, 17 June 2000, 48.

11. Shiva, *Biopiracy*, 120.

12. Shiva, *Biopiracy*, 121.

13. Robert Nozick, *Anarchy, State and Utopia* (New York: Basic, 1974).

14. Shiva, *Biopiracy*, 123–24.

15. "Italy to Scientists: No GM Gelato," Environmental News Service, 5 May 2000.

16. Nuffield Council on Bioethics, *Report on Genetically-Modified Food*, as reported in *London Daily Telegraph*, 28 May 1999.

17. Jeffrey Kluger, " Suicide Seeds," *Time*, www.acs.ucalgary.ca/~pubconf/Media/time.html.

18. *New Scientist*, 28 March 1998, www.hwcn.org/~ac096/articles/monsanto.html.

19. www.monsanto.com/monsanto/gurt/background/MIT.html.

20. *New Scientist*, 28 March 1998.

21. American Crop Protection Association Home web page, www.acpa/public/issues/index.html.

22. Kluger, "Suicide Seeds.

23. Ethirajan Anbarasan, "Dead-End Seeds Yield a Harvest of Revolt," *UNESCO Courier*, www.unesco.org/courier/1999_061/uk/ethique/txt1.htm

24. *New Scientist*, 28 March 1998.

25. www.monsanto.com/monsanto.

26. Harden, "Of Genes and Little Boys," *Washingtonian*, June 1984, 120.

27. Jane Rissler and Margaret Mellon, *The Ecological Risks of Engineered Crops* (Cambridge, Mass.: MIT Press, 1996), 29.

28. Carey Goldberg, "500 March in Boston to Protest Biotech Food," *New York Times*, 27 March 2000, Internet edition.

29. DeGregori, "Genetically Modified Nonsense."

CHAPTER 4

1. Review of *Killer Algae: The True Tale of a Biological Invasion*, by Alexander Meinesz, trans. D. Simberloff, *New York Times*, December 3, 1999, B4.

2. "French AIDS Scandal Verdict," Associated Press, 9 March 1999.

3. "Mad Cow and Human Prion Disease," *NeuroNews*, 18 May 1998. http://neuroscience.about.com/science/neuroscience/library/weekly/aa051898.htm.

4. Anne Maddocks, quoted in Antony Barnett and Patrick Wintour, "Revealed: Science Blunder That Gave Us BSE," *Guardian*, 8 August 1999.

5. Claire Ainsworth, "A Killer Is Born," *New Scientist*, 4 November 2000, 7.

6. Tim Radford, "Little Profit in Feeding Cattle Grass and Hay," *Guardian*, 28 October 1999.

7. "BSE Report: How It Went So Horribly Wrong," *New Scientist*, 4 November 2000, 4.

8. Antony Barnett, "Revealed: The Secret BSE Peril," *Guardian*, 22 August 1999.

9. John Colinge, "CJD Could Become an Epidemic of Biblical Proportions," *Times* (London), 7 August 1997.

10. Sarah Leyall, "Illness That Haunts Europe: One Family's Story," *New York Times*, 14 December 2000, A4.

11. Sandra Blakeslee, "Increase in Mad Cow Deaths Alarms Experts," *New York Times*, 25 July 2000, D1, 4.

12. Leyall, "Illness That Haunts Europe," A4.

13. James Meikle, "Drugs May Have Been Made from BSE-Infected Cattle," *Guardian*, 8 October, 1999, www.guardian.co.uk/bse/article/0.2763.388267.00.html.

14. James Miekle, "BSE Hearings End after Two Years," *Guardian*, 17 December 1999, www.guardian.co.uk/food/story/0.2763.195105.00.html.

15. Stanley Pruisner, "Shattuck Lecture—Neurodegenerative Diseases and Prions," *New England Journal of Medicine* 344, no. 20 (2001): 1516–26.

16. John Henley, "Beware the Food on Your Fork," *Guardian*, 30 August 1999, www.guardian.co.uk/food/story/0.2763.201176.00.html.

17. Raf Casert, "Belgium Admits Waste in Animal Feed," Associated Press, 19 September 1999.

18. "Dioxin Unlikely to Harm Belgian Health," *Reuters World Report*, 16 September 1999.

19. "PCBs and Human Health," U.S. Environmental Protection Agency, www.epa.gov.hudson/humanhealth/htm.

20. James L. Graff, "A Big Fizzle for Coca-Cola," *Business and Finance* 153, no. 25 (1999).

21. Alison Mutler, "EU Denounces Cyanide Spill, " Associated Press, 18 February 2000.

22. Jonathan Leake, "New BSE Outbreak Linked to Blood in Feed," *London Sunday Times*, 24 September 2000, A1, 14.

23. Eric Schlosser, *Fast Food Nation* (Boston: Houghton Mifflin, 2001), 202.

24. Caroline Smith DeWaal, testimony before Senate Commerce Committee subcommittee hearing on mad cow disease, 4 April 2001.

25. Jocelyn Gecker, "Human Cost of Mad Cow Disease," Associated Press, 29 November 2000 (*Birmingham News*, 8A).

26. Suzanne Daley, "Mad Cow Disease Spreads Faster Than the Disease," *New York Times*, 26 November 2000 (Internet edition).

27. G. Cannon, forward to Richard Lacey, *Mad Cow Disease*.

28. Michael Jacobs, *Guardian*, 7 July 1996, quoted in John Adams, "Cars, Cholera, and Cows: The Management of Risk and Uncertainty," *Policy Analysis* 335 (4 March 1999), www.cato.org/pubs/pas/pa-335es.html.

29. Vandana Shiva, *Stolen Harvest: The Hijacking of the Global Food Supply* (Cambridge, Mass.: South End, 2000), 72.

30. Cannon, forward to Lacey, quoted in Adams, "Cars, Cholera, and Cows."

31. "EU Bans Animal Products in Animal Feed," Associated Press, 5 December 2000 (*Birmingham News*, 8A).

CHAPTER 5

1. "Biotech Critics Cite Unapproved Corn in Taco Shells," *Washington Post*, 18 September 2000, A.

2. Phillip Brasher, "Biotech-Corn Product Recall Widened," Associated Press, *Birmingham News*, 14 October 2000, A7.

3. Paul Thompson, "Food Safety: Ethics and Agricultural Biotechnology," *Science of Food and Agriculture* 5, no. 1 (January 1993): 8.

4. Vandana Shiva, "Stolen Harvest: The Hijacking of the Global Food Supply" (Cambridge, Mass.: South End, 2000), 60.

5. Rifkin did not create these names, but merely popularized them once others did. One early reference to "Frankenfood" was a June 28, 1992, headline in the *New York Times,* "Geneticists' Latest Discovery: Public Fear of 'Frankenfood.'" The term had appeared a week earlier there in a letter to the editor (posting #51, AgBio archive, by Jay Byrne, on Monsanto Corporation), www.agbioworld.org.

6. Quoted in Blaine Harden, "Of Genes and Little Boys," *Washingtonian,* June 1984, 122.

7. Jeremy Rifkin and Robert L. Heilbroner, *The End of Work* (New York: Tarcher, 1996)

8. Jeremy Rifkin, *Beyond Beef: The Rise and Fall of the Cattle Culture* (New York: Plume, 1993).

9. Jeremy Rifkin, *The Emerging Order: God in the Age of Scarcity* (New York: Putnam, 1989).

10. Blaine Harden, "Of Genes and Little Boys," *Washingtonian,* June 1984, 189.

11. Leon Kass, "The Immortality Project," *First Things,* May 2001, 17–24.

12. Russ Hoyle, "Rifkin Resurgent," *Biotechnology* 10 (November 1992): 1406.

13. Jeremy Rifkin, *The Biotech Century* (New York: Putnam/Tarcher, 1998).

14. Hoyle, "Rifkin Resurgent," 1406.

15. Henry Miller, "The EPA's Bad Orange," *Washington Times,* 31 December 1998.

16. Hoyle, "Rifkin Resurgent," 1406.

17. Belinda Martineau, *First Fruit: The Creation of the FlavrSavr Tomato and the Birth of Biotech Food* (New York: McGraw-Hill, 2001).

18. "Mississippi Investigating Monsanto's Cotton," *Memphis Commercial Appeal,* 18 August 1997.

19. Patricia Reaney, Reuters, posted at *Eating Well Food News,* www.psa-rising.com/eatingwell/broccoli-super2000may.htm

20. Michael Brumas, "Gene-Altered Crops from State Facing Hostility," *Birmingham News*, 14 November 2000, A1.

21. J. O'Hara, "Food Allergies Complicate Diet," *Birmingham News*, 11 November 2000, B1.

22. "Food Allergy and Food Intolerance," *Mayo Clinic Family Health Book* (New York: Morrow, 1996), 1048.

23. "Food Allergy and Food Intolerance," *Mayo Clinic Family Health Book*, 1049.

24. Barnaby J. Feder, "Farmers Cite Scarce Data in Corn Mixing: Companies' Warnings Are Called Inadequate," *New York Times*, 17 October 2000.

25. "Genetic Scientist Suspended," BBC News, 12 August 1998.

26. "Is It or Isn't It? We're No Closer to Knowing If Genetically Modified Food Is Safe," *New Scientist*, 4 March 2000, 5.

27. Jan Brunvand, *Curses! Broiled Again: The Hottest Urban Legends Going* (New York: Norton, 1990).

28. Lindsay Bond Totten, "Biggest Impact? Genetic Engineering," in the Garden column, Scripps Howard News Service, printed in the *Birmingham News*, 30 December 1999, B1, B3.

29. For example, see the Gentech archive: www.gene.ch/gentech/1999/Jan-Feb/msg00150.html.

30. R. Paul Ross, "Impact of Genetically Modified Food on the Dairy Industry: An Irish Perspective," 1999. This is an article distributed at a dairy conference in 1999. It was supplied to me by Mary Ellen Sanders, dairy consultant (see acknowledgments).

31. Richard E. Goodman, "Genetically Engineered Crops: How? Why? Are They Safe?" (lecture at the University of Alabama at Birmingham, sponsored by the Microbiology Department, May 24, 2000).

32. John Carey, "Are BioFoods Safe?" *Business Week*, 20 December, 1999, 72.

33. Nicols Fox, *Spoiled: Why Our Food Is Making Us Sick and What We Can Do about It* (New York: Basic, 1997), 93.

34. Bill Grierson, *American Council on Science and Health* (ACSH) "Food Safety through the Ages," 9, no. 3 (1997), www.acsh.org.publications/priorities/0903/foodsafety/html/.

35. Fox, *Spoiled,* 92.

36. Rebecca Goldberg, quoted in "Seeds of Change," *Consumers Reports,* September 1999, 42.

37. Ellen Goodman, "Know Where Fast Food Comes From?" *Boston Globe,* 21 February 2001.

38. Candy Sagon, "Fast Food's Foe," *Washington Post,* 28 February 2001, F 4.

39. Fox, *Spoiled,* 238.

40. Eric Schlosser, *Fast Food Nation* (Boston: Houghton Mifflin, 2001), 150.

41. Quoted in Fox, *Spoiled,* 252.

42. Lance Gay, "Graphic Video Causes Outrage over Slaughterhouse Practices," *Birmingham Post-Herald,* 31 January 2001, A6.

43. Schlosser, *Fast Food Nation,* 150.

44. Schlosser, *Fast Food Nation,* 139, 150.

45. Office of Media Communication, CDC, "CDC Data Provides the Most Complete Estimate on Foodborne Disease in the United States," www.cdc.gov/od/oc/media/pressrel/r990917.htm.

46. Schlosser, *Fast Food Nation,* 197.

47. Schlosser, *Fast Food Nation,* 205–6.

48. Schlosser, *Fast Food Nation,* 211–12, 210.

49. Schlosser, *Fast Food Nation,* 211–12.

50. Schlosser, *Fast Food Nation,* 211–12, 218.

51. An anonymous "slaughterhouse engineer," quoted in Schlosser, *Fast Food Nation*, 211–12, 218.

52. Steven Bjerklie, quoted in Schlosser, *Fast Food Nation*, 200, 218.

53. Schlosser, *Fast Food Nation*, 125–26.

54. Edward Groth, III, "Consumer's Union Calls EPA . . . Inadequate," press release, 2 August 1999, www.consumersunion.org/food/epadc899.htm.

55. Robert Pear, "Tighter Rules Are Sought for Dietary Supplements," *New York Times*, 17 April 2001.

56. "'Mad Cow Disease': Where Do We Go from Here?" *Consumer Reports*, May 2001, 6.

57. "Like Lambs to the Slaughter: What If You Can Catch Old-Fashioned CJD by Eating Meat from a Sheep Infected by Scrapie?" *New Scientist*, 31 March 2001, 4.

58. "A Consumer's Guide to Pesticides and Food Safety," Cooperative Extension College of Agriculture and Food Sciences, University of Arizona, November 1995, www.ag.arizona.edu/pubs/health/foodsafety/921079.html.

59. David Barboza, "Gene-Altered Corn Changes Dynamics of Grain Industry," *New York Times*, 9 December 2000.

CHAPTER 6

1. Ernst Lehmann, *Biologischer Wille. Wege und Ziele biologischer Arbeit im neuen Reich* (Munich, 1934), 10–11; quoted in Peter Staudenmaier, "Fascist Ecology: The 'Green Wing' of the Nazi Party and Its Historical Antecedents," *Ecofascism: Lessons from the German Experience*, ed. Janet Biehl and Peter Staudenmaier (San Francisco: AK Press, 1995).

2. Jean-Jacques Rousseau, *Confessions*, ed. Patrick Coleman, trans. Andrea Scholar (Oxford: Oxford University Press, 2000); Rousseau, *Emile*, ed. P. D. Jimack (Everyman's Library, 1993).

3. Lawrence Buell, "Thoreau and the Natural Environment," in *The Cambridge Companion to Thoreau,* ed. Joel Myerson (New York: Cambridge University Press, 1995), 171.

4. Ralph Waldo Emerson, *Journals of Ralph Waldo Emerson, with Annotations*, ed. Ralph Waldo Emerson and Waldo Emerson Forbes (1820–1824; reprint, Boston: Houghton Mifflin, 1909), 354–56.

5. Mindy Sink, "Radicals Take Responsibility for Burning Timber Office," *New York Times,* 11 January 2001.

6. Ralph Waldo Emerson, *The Journals and Miscellaneous Notebooks of Ralph Waldo Emerson,* ed. William H. Gilman et al., 16 vols. (Cambridge: Harvard University Press, 1960–1982), 14:203; quoted in Robert Sattelmeyer, "Thoreau and Emerson," in *The Cambridge Companion to Thoreau,* 35.

7. Henry David Thoreau, *The Writings of Henry David Thoreau: Journal* (1981–), ed. John C. Brederick et al., vol. 4 (Princeton, N.J.: Princeton University Press, 1992), 137 (October 10, 1851); quoted in Robert Sattelmeyer, "Thoreau and Emerson," in *The Cambridge Companion to Thoreau,* 35.

8. Joseph Wood Krutch, *Henry David Thoreau* (New York: Sloane, 1948), 5.

9. Quoted in Ronald Bailey, *Eco-Scam: The False Prophets of Ecological Apocalypse* (New York: St. Martin's, 1993), 16, quoting Riley Dunlap, "Public Opinion in the 1980s: Clear Consensus, Ambiguous Commitment," *Environment* (October 1991): 32.

10. William James, *Varieties of Religious Experience* (New York: Longman Green, 1925), 150.

11. Stephen Toulmin, *The Return to Cosmology: Postmodern Science and the Theology of Nature* (Berkeley: University of California Press, 1982), 271.

12. David Hume, *Dialogues Concerning Natural Religion,* ed. Henry David Aiken (New York: Hafner, 1948, 1972), 79.

13. Tom Regan, *The Case of Animal Rights* (Berkeley: University of California Press, 1985).

14. Holmes Rolston III, *Environmental Ethics: Duties to and Values in the Natural World* (Philadelphia: Temple University Press, 1988).

15. http://lamar.colostate.edu/~rolston/grad-csu.htm.

16. Aldo Leopold, *A Sand County Almanac, Robert Finch (Introduction)* (reprint ed.; New York: Oxford University Press, 1987).

17. "Are Human Beings a Cancer on the Biosphere?" *Proceedings and Addresses of the APA* 74, no. 4 (2001).

18. Vandana Shiva, *Biopiracy: The Plunder of Nature and Knowledge* (Boston: South End, 1997).

19. Carolyn Merchant, *Radical Ecology: The Search for a Livable Ecology: Revolutionary Thoughts/Radical Movements* (London: Routledge, 1992).

20. Mary Daly, *Gyn-Ecology: The Metaethics of Radical Feminism* (Boston: Beacon, 1978).

21. Arne Naess, "Identification As a Source of Deep Ecological Attitudes," in *Deep Ecology*, ed. Michael Tobias (San Diego: Avon, 1985), 260.

22. I am indebted to Louis Pojman's crystallization of Taylor's argument in *Global Environmental Ethics* (Mountain View, Calif.: Mayfield, 1999), 188.

23. Freya Mathews, "Value in Nature and Meaning in Life," *Environmental Ethics*, ed. Robert Elliot (Oxford: Oxford University Press, 1992), 151.

24. Bill Devall and George Sessions, *Deep Ecology: Living As If Nature Mattered*, Salt Lake City, Peregrine Press, 1085. Quoted in Pojman, *Global Environmental Ethics*, 179.

25. Arthur Danto, *Mysticism and Morality* (New York: Basic Books, 1972).

26. Paul Johnson, *Intellectuals* (New York: Harper Perennial, 1988).

27. Wendell Berry, *The Unsettling of America: Culture and Agriculture*, (San Francisco: Sierra Club Books, 1986), 29.

28. Wendell Berry, *The Unsettling of America*, 29.

29. Staudenmaier, "Fascist Ecology."

30. Michael Zimmerman, *Heidegger's Confrontation with Modernity: Technology, Politics, and Art* (Indianapolis: Indiana University Press, 1990), 242–43; quoted in Staudenmaier, "Fascist Ecology," 12.

31. Peter Staudenmaier, "Fascist Ecology," 10.

32. Reproduced in Joachim Wolschke-Bulmahn, *Auf der Suche nach Arkadien* (Munich, 1990), 147; quoted in Peter Staudenmaier, "Fascist Ecology," 10.

33. Peter Staudenmaier, "Fascist Ecology," 18, 20, 22.

34. Peter Staudenmaier, "Fascist Ecology," 19.

35. Ronald Bailey, *Eco-Scam: The False Prophets of Ecological Apocalypse* (New York: St. Martin's, 1993), 16.

36. Gregory Bassham, review of *Environmental Ethics: Concepts, Policy, Theory*, by Joseph DesJardins, *Teaching Philosophy* 4 (March 2001): 86. By the way, the book under review does *not* fall under the reviewer's description of most such textbooks.

CHAPTER 7

1. R. R. Palmer and Joel Colton, *A History of the Modern World* (New York: Knopf, 1969), 431–32.

2. Garrett Hardin, "The Tragedy of the Commons," *Science* 162 (1968): 1243–48.

3. William Paddock and Paul Paddock, *Famine 1975! America's Decision: Who Will Survive?* (Boston: Little, Brown, 1967).

4. Garrett Hardin, "Carrying Capacity As an Ethical Concept," in *Lifeboat Ethics: The Moral Dilemmas of World Hunger*, ed. G. Lucas and T. Ogletree (New York: Harper Forum Books, 1976).

5. Garrett Hardin, "Carrying Capacity As an Ethical Concept," 126.

6. Garrett Hardin, "Carrying Capacity As an Ethical Concept," 132.

7. Garrett Hardin, "Carrying Capacity As an Ethical Concept," 123, 129.

8. Peter Rosset, "Why Genetically Altered Food Won't Conquer Hunger," *New York Times*, 1 September 1999.

9. Advertisement, *New York Times*, 18 January 2000, A13.

10. Amartya Sen, *Poverty and Famines: An Essay on Entitlement and Deprivation* (New York: Oxford University Press, 1981), 160.

11. Vandana Shiva, quoted by Noel Holton, in "Biotech Aimed at World Hunger," *News*, 29 June 2000.

12. Kelvin Nig, "'Golden Rice' Has No Shine, Say Critics," Third World Network (TWN), www.twnside.orgn.sg/title/golden.htm.

13. Kelvin Nig, "'Golden Rice' Has No Shine, Say Critics."

14. Kelvin Nig, "'Golden Rice' Has No Shine, Say Critics."

15. Norman Borlaug, quoted in "The Life and Work of Norman Borlaug, Nobel Laureate," by Robert W. Herdt, 14 January 1998, reproduced at http://208.240.21/agsci/robertherdt.html.

16. Herdt, "The Life and Work of Norman Borlaug."

17. Gregg Easterbrook, "Forgotten Benefactor of Humanity," *Atlantic Monthly* 270, no. 1 (January 1997): 75–82.

18. Garrett Hardin, "Lifeboat Ethics: The Case against Helping the Poor," *Psychology Today* (September 1974). Reprinted in W. Aiken and H. LaFollette, *World Hunger and Moral Obligation* (Englewood Cliffs, N.J.: Prentice-Hall, 1977), 18.

19. Easterbrook, "Forgotten Benefactor of Humanity," 75–82.

20. Easterbrook, "Forgotten Benefactor of Humanity," 75–82.

21. Herdt, "The Life and Work of Norman Borlaug."

22. Norman Borlaug, "Billions Served," interview by Ronald Bailey, *Reason*, April 2000, 36.

23. Bill Gates, "Will Frankenfood Save the World?" *Time*, 13 June 2000.

24. Ronald Bailey, "Billions Served," *Reason*, April 2000, 36.

25. Florence Wambugu, "Feeding Africa," interview, *New Scientist*, 27 May 2000, 40–43.

26. "Seven Academies of Science Urge Action to Promote Use of Biotech in Alleviating World Hunger, Poverty," National Academies, 12 July 2000, www4.nationalacademies.org/news.nsf/.

CHAPTER 8

1. Mark Sagoff, "What's Wrong with Exotic Species?" *Reason*, March 2000.

2. Mark Sagoff, "What's Wrong with Exotic Species?"

3. Steven J. Milloy, "Environmentalists and Killer Bees," *San Diego Union Tribune*, 23 June 2000.

4. "Bt Corn and European Corn Borer: Long-Term Success through Resistance Management," University of Michigan Extension Service, 1997, at www.extension.umn.ledu/distribution/cropsystems.DC7055.html

5. Gary Munkvold, "Disease Control with Bt Corn," Integrated Crop Management, Department of Entomology, Iowa State University, Ames, Iowa, www.ent.iastate.edu/ipm/icm/1998/1-19-1998/btdiscon.html.

6. John Losey et al., "Transgenic Pollen Harms Monarch Larvae," *Nature* 399 (20 May 1999): 214.

7. *New York Times*.

8. "Toxic Pollen from Widely-Planted, Genetically Modified Corn Can Kill Monarch Butterflies, Cornell Study Shows," Cornell University news release, 19 May 1999, www.news.cornell.edu.releases/May99/Butterflies.bpt.html.

9. "To Bt or Not to Bt: The Sound Science That Brought Down Bt Crops," Third World Network, www.twnside.org/sg.title.crops.htm.

10. Marc Lappé and Britt Bailey, *Against the Grain: Biotechnology and the Corporate Takeover of Your Food* (Monroe, Me.: Common Courage, 1998), 97.

11. Marc Lappé and Britt Bailey, *Against the Grain,* 100.

12. Sarad Joshi, "An Indian Farmer Rebutts Vandana Shiva," AbBioView home page, 26 May 2000, Archive message #90, www.agbioworld.org.

13. Stephen R. Palumbi, "The High-Stakes Battle over Brute-Force Genetic Engineering," *Chronicle of Higher Education,* 13 April 2001, B7–9.

14. Marc Wortman, "Medicine Gets Personal," *Technology Review,* January–February 2001, 78.

15. Jane Rissler and Margaret Mellon, *The Ecological Risks of Engineered Crops* (Cambridge, Mass.: MIT Press, 1996), 2.

16. Marc Lappé and Britt Bailey, *Against the Grain,* 97.

17. Rissler and Mellon, *The Ecological Risks of Engineered Crops,* 6.

18. "Modified Crops Could Corrupt Weedy Cousins," *New Scientist,* 15 July 2000, 6.

19. "Modified Crops Could Corrupt Weedy Cousins," 6.

20. Rissler and Mellon, *The Ecological Risks of Engineered Crops,* 30.

21. City of Santa Monica, "Santa Monica Municipal Code," http://pen2.ci.santa-monica.ca.us/city/municode/codemaster/Article_7/44/010.html.

22. Bryan J. Hubbell and Rick Welsh, "Transgenic Crops: Engineering a More Sustainable Agriculture?" *Agriculture and Human Values* 15 (1998): 43–56; Sindel (1996) is B. Sindel, "Glyphosate Resistance Discovered in Annual Ryegrass," *Resistant Pest Management* 8 (1996): 5-6; Gressel (1996) is J. Gressel, "Fewer Constraints Than Proclaimed to the Evolution of Glyphosate-Resistant Weeds," *Resistant Pest Management* 8 (1996): 2–5; and Monsanto (1997) is "Responses to Questions Raised and Statements Made by Environmental/Consumer Groups and Other Critics of Biotechnology and Roundup Ready Soybeans," Monsanto Crossroads, Monsanto Corporation, 21 April 1997, www.monsanto.com.

23. Hubbell and Welsh, "Transgenic Crops," 47.

24. Rissler and Mellon, *The Ecological Risks of Engineered Crops.*

25. Hubbell and Welsh, "Transgenic Crops," 47.

26. Rissler and Mellon, *The Ecological Risks of Engineered Crops,* 43.

27. Marc Lappé and Britt Bailey, *Against the Grain,* 96.

28. www.monsanto.com/monsanto.

29. Laura Hansen and John Obrycki, "Non-target Effects of Bt Corn Pollen on the Monarch Butterfly," www.ent.iastate.edu/entsoc/ncb99/prog/abs/D81.html

30. Ken Hough, "Updates on Bt Corn Research," Ontario Corn Producers Association, www.ontariocorn.org/Jan2000rr.

31. Marlin Rice, "Monarchs and Bt Corn: Questions and Answers," Iowa State University, 14 June 1999, www.ext.iastate.edu/ipm/icm/1999/6-14-99/monarchbt.html.

32. Mark Sears, University of Guelph.

33. Ken Hough, "Updates on Bt Corn Research," Ontario Corn Producers Association, www.ontariocorn.org/Jan2000rr.

34. Carol Kacsuk Moon, "Modified-Crop Studies Are Called Inconclusive," *New York Times,* 14 December 2000, A31.

35. "EPA Sets New Restriction on Genetically Altered Corn," Associated Press, *Birmingham News,* 17 January 2000, 3A.

36. Mae-Wan Ho and Angela Ryan, "Swallowing the Tale of the Swallowtails: No 'Absence of Toxicity' of Bt Pollen," Third World Network, www.twnside.org.sg.title.swallow.

CHAPTER 9

1. Wendell Berry, *The Unsettling of America: Culture and Agriculture* (San Francisco: Sierra Club Books, 1986), 53.

2. Wendell Berry, *The Unsettling of America*, 843. As an antidote to Barry, see Wallace Kaufman, *Coming Out of the Woods* (Cambridge, Mass.: Perseus, 2000).

3. Quoted in Wendell Berry, *The Unsettling of America*, 8.

4. Jared Diamond, *Guns, Germs, and Steel* (New York: Norton, 1997).

5. Ronald Bailey, "Rage against the Machines," *Reason*, July 2001, 31.

Index

About the Author

Gregory Pence has taught bioethics for more than twenty-five years in the Philosophy Department and School of Medicine at the University of Alabama at Birmingham, where he has won the Best Teacher award. He has lectured in more than two hundred universities worldwide and published in the *New York Times*, *Wall Street Journal*, and *Newsweek*. He has published *Who's Afraid of Human Cloning?* and *Re-Creating Medicine* and edited *Flesh of My Flesh: The Ethics of Human Cloning* and *The Ethics of Food: A Reader for the Twenty-first Century* (all with Rowman & Littlefield). His text with McGraw-Hill *Classic Cases in Medical Ethics* will soon go into its fourth edition.